The Molecular Basis of Sex and Differentiation

Milton H. Saier
Gary R. Jacobson

The Molecular Basis of Sex and Differentiation

A Comparative Study of Evolution, Mechanism, and Control in Microorganisms

With 100 Illustrations

Springer-Verlag
New York Berlin Heidelberg Tokyo

Milton H. Saier Jr.
Department of Biology
The John Muir College
University of California
  at San Diego
La Jolla, CA 92093
U.S.A.

Gary R. Jacobson
Department of Biology
Boston University
Boston, MA 02215
U.S.A.

QR
84
.S326
1984

Library of Congress Cataloging in Publication Data
Saier, Milton H.
    The molecular basis of sex and differentiation.
    Bibliography: p.
    Includes index.
    1. Micro-organisms—Physiology. 2. Microbial
differentiation. 3. Sex (Biology) 4. Cellular control
mechanisms. I. Jacobson, Gary R. II. Title.
QR84.S326   1984   576'.11   84-10905

Cover: Yeast sexual agglutination factors (see Figure 10.4, page 142).

Media conversion by WorldComp, Inc., New York, New York.
Printed and bound by Halliday Lithograph, West Hanover, Massachusetts.
Printed in the United States of America.

9 8 7 6 5 4 3 2 1

ISBN 0-387-96007-4 Springer-Verlag New York Berlin Heidelberg Tokyo
ISBN 3-540-96007-4 Springer-Verlag Berlin Heidelberg New York Tokyo

for Jeanne

Preface

Man's mind stretched to a new idea
never goes back to its original dimensions
Oliver Wendell Holmes

Our current understanding of sex and biological differentiation results from the application of three principal experimental approaches to these subjects: those of the physiologist, the biochemist, and the geneticist. These three approaches are illustrated by the materials presented in the chapters of this volume. Chapters 1–5 emphasize conceptualization of developmental processes, describing systems principally from the standpoint of the physiologist. Structures and functions are defined with only occasional reference to specific molecular details. Chapters 6–10 present the views of the biochemist, attempting to describe functions influencing or regulating cellular behavior at the molecular level. And Chapters 11–14 illustrate the approaches of the modern-day geneticist in his attempts to gain a detailed understanding of processes controlling gene expression.

While it is possible to delineate these three major sections, each emphasizing a distinct experimental approach, it must be realized that the yield of knowledge increases exponentially with the number of experimental approaches available to the investigator. Information resulting from the application of each of these approaches must converge to give the same answers for any one biological phenomenon in any one experimental system. Further, if we can learn of details regarding a particular process by applying different experimental approaches, our postulates concerning the underlying molecular mechanisms are likely to be more accurate.

But biological systems are not unrelated. Evolution provides the link that allows us to relate processes occurring in one organism to those operative in all others. A function is likely to be similarly performed in organisms that diverged relatively late in evolutionary history, while those diverging earlier will exhibit greater differences. With evolutionarily divergent organisms, mechanistic differences may have evolved to accommodate the differing degrees of complexity of cellular construction or to coordinate functions of differentiated cells in a multicellular organism. Still, we must realize that the basic life-endowing molecular processes had to exist prior to extensive evolutionary divergence—even before the appearance of two distinct cell types. Consequently, we should expect that these processes are governed by the same principles, and that even the molecular details will sometimes be conserved throughout evolutionary history. Based on these postulates, it can be assumed that information available through the study of microorganisms will be directly applicable to higher organisms, and that molecular processes controlling complex interactions in multicellular organisms will have their rudiments in the essential life-endowing characteristics of the simplest bacteria.

This "unity principle in biology" is possibly the most important concept facing the modern-day biologist. It allows extrapolation of information from one organism to another as well as integration of basic facts obtained through the study of multiple species. Nevertheless, it does not preclude diversity. A given function may be performed by several distinct mechanisms, and a particular organism may have selected one such mechanism while another has chosen a second. Moreover, an organism may perform a function in more than a single way, and the two "pathways" may compete with, complement, or be superimposed on each other. Multiple pathways allow for fine control and permit more rapid evolutionary change.

This unifying maxim is tacitly applied by most developmental biologists. It is widely accepted that morphogenetic forces operative during embryogenesis (Chapter 1) will exhibit features common to those of developing bacteria and eukaryotic microorganisms (Chapters 2 and 5). The concepts of cellular and organismal mortality, to be contrasted with the potential for immortality in higher organisms (Chapter 3), have been formulated through studies with microorganisms (Chapters 2, 4, and 13). The similarities noted in Chapter 4 among cyclic programs of differentiation in prokaryotic and eukaryotic microorganisms and the molecular mechanisms underlying these programs (Chapter 5) also attest to a universal evolutionary origin.

As discussed in Chapter 6, microorganisms control their internal environment by restricting or facilitating the stereospecific movement of molecules across their membranes. Multiple mechanisms of translocation have evolved and most of these occur universally throughout the living world. Similarly, micromolecular and macromolecular reception processes as well as intercellular recognition (Chapters 7–10) are likely to be mediated by functionally analogous cell surface proteins and glycoproteins. These reception processes must trigger intracellular transmission mechanisms (Chapters 7 and 9), which elicit appropriate biological responses. Finally, the genetic material of the cell must be differentially articulated, depending on the stage of differentiation in which the cell finds itself. Owing in large part to the advent of rapid gene cloning and DNA sequencing techniques, the molecular details of genetic regulatory processes are now coming to light (Chapters 11–13). Again, our appreciation of the diversity of genetic regulatory mechanisms operative in the biological world relies upon the multiorganismal approach to the problem.

The present monograph is designed to familiarize the reader with the essential unifying concepts in developmental biology. In attempting to illustrate these

principles we shall wander to the edges of (and beyond) the frontiers of our
knowledge. When possible, these concepts will be pursued to the molecular
level. Extensive reference to experimental detail has been omitted in order to
maximize conceptual recognition of the underlying principles governing cellular
and organismal development. Only when a knowledge of experimental protocol
is essential to understanding the process under discussion will this information
be presented. Selected references at the end of each chapter are provided to allow
the reader to pursue a subject in greater depth.

> What is true for *Escherichia coli*
> is true for elephants, only more so.
>
> Jacques Monod

Acknowledgments

We would like to thank our colleagues and friends who contributed to the formulation of this volume:

Clint Ballou, Sam Barondes, Hans Bode, Stu Brody, Willie Brown, Adelaide Carpenter, Gary Cote, Tom Dallman, Ted DeFrank, Steve Dills, Elis Englesberg, Dave Epel, Keith Fischer, George Fortes, Rob Hausman, Dale Kaiser, Pat Lakin-Thomas, Bill Loomis, Michael Newman, Michael Novotny, Ed Orias, Sheila Podell, Lola Reid, Mark Rose, Lucy Shapiro, Mel Simon, Meredith Somero, Jeff Stock, Jeremy Thorner, Hoyt Yee.

Particular thanks are extended to Pat Gifford who drew the illustrations and to Carmen Jacobo and Leanne Sorensen who provided excellent secretarial assistance in the preparation of this manuscript.

Contents

CHAPTER 14
Conclusions and Perspectives

Index

CHAPTER 1

Embryology and the Study of Microbial Development

The study of embryology, and consequently of biological development, was initiated by Aristotle in 340 B.C. He followed and recorded the development of a chick embryo within the egg, noting that the developing embryo went through distinct chronological stages that resembled those of other organisms. From this observation stemmed the postulate that ontogeny is a recapitulation of phylogeny (Baer's Law).

Two principal antagonistic theories were put forth to account for embryological development: Preformation and Epigenesis. The Preformation theory stated that a preformed creature resembling the adult organism was present in the embryo throughout development; the epigenesis theory suggested the emergence of morphological features during development. With the discovery of sperm in the 1700s, the preformationists assumed that since the father contributed the "Creative Principle," the "little man" had to be contained within the sperm cell. The discovery of parthenogenesis in 1750 and the dissection of early embryos in 1760, revealing only fluid and granuoles (no little man), were severe blows to the preformation hypothesis.

In 1888, Roux, the father of experimental embryology, allowed a fertilized frog egg to divide, killed one of the two daughter cells with a hot needle, and observed the consequences. The morphologically defective embryo that developed resembled half of a tadpole. Preformation appeared to be supported. Roux did not attempt to remove the dead cell from the live one, and it was not until 1930 that the experiment was repeated with cells separated from 2-cell stage embryos. Each cell gave rise to a normal tadpole, thus dealing the final blow to preformation.

In the 1950s, Gurdon carried out nuclear transplantation experiments in which nuclei from differentiated frog cells were transferred to anucleate egg cells. A percentage of the transplanted eggs developed into viable tadpoles, showing that the master plan for development must be contained within the nucleus. This result and subsequent experiments have led to two major conclusions: (1) that differential gene expression controls differentiation, and (2) that gene expression must be subject to extranuclear control.

Observational studies with the polytene chromosomes of *Drosophila* first led to the concept of genetic programs. The formation of chromosomal "puffs" during early pupation, attributable to RNA synthesis from temporally regulated genes, was studied. Several major conclusions resulted from this work: (1) for any given tissue, puffing occurred at a reproducible time during development; (2) the puffing pattern was tissue specific, and each tissue exhibited its characteristic pattern; (3) the puffing patterns during pupation were reversibly controlled by two antagonistic hormones, the pupation hormone (ecdysone) and the juvenile hormone. Administration of ecdysone gave rise to pupation puffs, while administration of the juvenile hormone gave rise to larval puffs. Thus, hormonal control of reversible gene expression was demonstrated at this stage in the developmental process.

Two possibilities were considered: Independently regulated gene products could control sequential gene expression, or there might be an obligatory sequence of genetic activation steps, such as

$$a \rightarrow b \rightarrow c \rightarrow d \rightarrow e \rightarrow f \ldots$$

In order to test these hypotheses, portions of the *Drosophila* chromosome were deleted (i.e., genes *b, c,* and *d*) leaving genes *a, e,* and *f* intact. The temporal regulation of the remaining genes was not altered. This observation favored the suggestion that transcriptional regulatory proteins controlled the temporal expression of at least some structural genes during development.

Studies with *Drosophila* also led to the realization that the expression of other genes led to irreversible developmental decisions. Further, expression of reversibly controlled genes may depend on the expression of an irreversibly controlled event. For example, once the bithorax gene is activated, the genes in the metathoracic segment are switched into a new pathway that produces wings. All "wing" genes are activated sequentially. It is now clear that gene activation can occur either reversibly or irreversibly, depending on the specific gene under study, and that the *irreversible* events represent important decision-making steps (*commitments*) occurring during embryogenesis.

Biochemical programs, clearly the direct result of the expression of genetic programs, have been characterized. Thus, synthesis of histone types during sea urchin development is temporally regulated (Figure 1.1). Some histones are synthesized early, others late in development, and none of the late-appearing histones are apparently derived from the earlier ones. The temporal sequence evidently involves activation and silencing of histone genes during specific stages in development.

Biologists concerned with developmental problems in biology have usually approached their systems from one of several standpoints. First, they have characterized the sequence of events that comprises a cyclic or linear progression

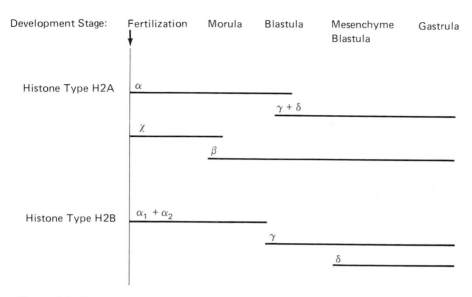

Figure 1.1. Temporal patterns of histone synthesis during development of a sea urchin embryo following fertilization. Two histone classes, types H2A and H2B are shown. The Greek letters refer to the histone subtype. (After L.H. Cohen, K.M. Newrock, and A. Zweidler, *Science 190:*994–997 (1975).)

using classical observational techniques or biochemical approaches. Second, they have examined the genetic mechanisms that allow expression of differentiation-specific functions. These genetic mechanisms can be divided into two classes: (1) those mechanisms which activate silent genes to a state in which expression becomes possible, and (2) those mechanisms that allow expression of an activated gene. Some of the former events may prove to be due to changes in the state of DNA methylation or to genetic rearrangements that render a gene *potentially* active (see Chapter 7). The latter events may correspond to regulatory interactions of proteins or other biological macromolecules with the nucleic acid resulting in enhanced rates of transcription and translation. Both genetic activation mechanisms may be required for expression of a particular differentiation-specific trait. Third, biochemists have examined individual processes peculiar to a differentiated cell in order to understand the mechanisms by which the aforementioned gene products function. Examples include mechanisms of cell–cell interaction (Chapter 10) and ligand–receptor interaction (Chapters 6–9), and these interactions may in turn influence gene expression. Some of these interactions may function to sensitize the genetic apparatus to concentrations of small molecules in the cytoplasm or extracellular matrix. Thus, feedback and feedforward regulatory mechanisms in gene control during development can be expected to be mediated by macromolecular receptors in the cytoplasm and plasma membrane.

In this volume we shall be concerned largely with three phenomena: Differentiation, Sex, and Death. It can be argued that these three topics are intimately and obligatorily related and, therefore, must be considered together. The advent of irreversibly developing organisms had to be accompanied by the appearance of programmed death and sexual reproductive mechanisms. In fact, the terminal step in any sequence of irreversible steps of cellular development must either be death or immortality, and few examples of linear differentiation leading to natural immortality are known. The suggestion that death is usually the terminal step in a sequence of irreversible differentiation steps results in part from the fact that no organism carries an infinite quantity of DNA, and this fact renders finite the number of steps through which the organism can pass. Moreover, if the mortal organism is to survive for more than one generation as a species, some rejuvenating principle must be available to it. The rejuvenating principle employed by most organisms is sexual conjugation, and conjugation gives rise to offspring of genetic compositions determined by the genetic complements of the parents. The biological phenomena of Differentiation, Sex, and Death are inextricably linked in the existence of any mortal organism.

In order to avoid ambiguity and to progress toward an understanding of biological development, a number of terms must be defined. *Evolution* is a progressive process that can give rise to a present day adult organism. One can consider *phylogenetic evolution,* in which case one refers to the sequential events that occurred throughout phylogeny, i.e., over biological evolutionary history on earth (about 3 billion years), during which the organism developed increasing degrees of complexity. Alternatively, we may consider *ontogenetic evolution,* referring to the development of a single cell or organism through ontogeny, from a fertilized egg to the last irreversible differentiative step.

Differentiation can be defined as the development of function, while *morphogenesis* refers to the development of shape. *Development* encompasses both differentiation and morphogenesis. It is perhaps best defined as "genetically directed changes in a single cell, or groups of cells, assumed in a programmed fashion over time."

Some organisms undergo irreversible but *cyclic development,* while others are only capable of *linear development.* For example, certain bacteria and lower eukaryotes such as yeast differentiate into resting spores which subsequently germinate into the vegetative organism. These organisms are said to cycle through development. By contrast, all "mortal" multicellular organisms pass through a linear program of differentiation, never cycling back to an earlier state. In order for such an organism to become *rejuvenated,* it must undergo *sexual conjugation* during which genetic material is transferred between organisms of a single species, giving rise to a new, genetically unique individual.

Death can be considered to be the cessation of life, but *life* is not so easily defined. We can compare *genetic, organismal,* and *cellular death.* A living organism that is incapable of reproduction is said to be genetically dead. By contrast, an animal that lacks a heartbeat and no longer emits brain waves is organismally dead, even though individual cells within its body may live on. Moreover, we can distinguish *accidental death* from *programmed death.* Accidental death refers to the loss of life as a result of environmental factors external to the organism itself. By contrast, programmed death refers to a terminal step in a program of irreversible differentiation that results in loss of life. Extending this argument one step further, we shall define a *mortal cell* or organism as one which is programmed to die while an *immortal cell* is one which can live forever, provided that external conditions remain favorable. By this definition, simple bacteria and germ cells of higher organisms must be considered to be immortal because they possess the potential to live forever. Unicellular ciliates, such as Paramecia, and somatic cells of higher organisms, on the other hand, are considered to be mortal because they are destined (programmed) to die. Only genetic changes that disrupt a developmental program can rescue an irreversibly differentiating cell from the ultimate fate of mortality.

Finally, what is meant by the term *microorganism*? This term has been defined, on structural grounds, as an organism which exists in the unicellular state; on functional grounds, as one which is easily subjected to physiological and genetic manipulation, and quantitatively, as one which is small. All of these definitions are arbitrary. Many microbes of both prokaryotic and eukaryotic cell structure can exist either in a unicellular or multicellular state depending on the stage of their life cycle. Technical definitions are difficult as techniques for the genetic and physiological manipulation of higher organisms are constantly becoming more refined. And most organisms would be considered small relative to an elephant or a redwood tree. For the purpose of this discussion, we shall define a *microorganism* as one which can be subjected to the experimental techniques of the microbiologist and can be reduced to the unicellular state for examination by the cell biologist. All the organisms to be discussed in this volume fall within this category.

Comparative studies of prokaryotic and eukaryotic microorganisms led to the suggestion that genetic mechanisms responsible for the various modes of differentiation may have evolved prior to the advent of nucleated cells. Thus, the numerous and striking similarities in developmental processes occurring in select members of the prokaryotic and eukaryotic worlds suggest that in divergent organisms, evolution has occurred with retention of a particular mode of development. A second possibility is that convergent evolution of developmental processes has occurred in divergent species, as a result of a common environment and similar evolutionary pressure. It should be pointed out that divergent and convergent evolution are not mutually exclusive, but could have occurred simultaneously. Common environmental pressures would be expected to promote the expression and utilization of shared genetic material of value within a particular environmental framework. Additionally, genetic exchange between evolving prokaryotes and eukaryotes undoubtedly occurred, and such exchange would be most likely between organisms sharing a common habitat. Numerous examples of amino acid sequence homology in proteins of eukaryotic and prokaryotic origin support these contentions. The thesis that convergent and divergent evolution represent components of a single process will be reiterated and discussed in Chapter 4. In the remainder of this chapter, several examples of superficial similarities between prokaryotes and eukaryotes will be discussed. Common molecular mechanisms and developmental programs will undoubtedly be shown to underlie some but not all of these similarities.

Certain bacteria, like animal cells, can alter their shape and interact with each other by recognition processes that must involve cell surface homotypic or heterotypic adhesive macromolecules (see Chapter 10). *Ancalomicrobium adetum* is a bacterium with a flexible star-shaped appearance that exhibits adhesive self-recognition properties, allowing for intimate contact. *Myxobacteria* also show different responses when encountering bacteria of the same and different species. Thousands of myxobacteria migrate in cohesive packs seeking live prokaryotic food sources which they lyse and devour. *Bdellovibrio,* on the other hand, is a small bacterium which does not adhere to like cells but will bind tenaciously to the outer membranes of certain Gram-negative bacterial species which it can parasitize and kill. Heterotypic adhesion induces the secretion of digestive enzymes which locally rupture the outer membrane and peptidoglycan layer of the larger prey bacterium, thus allowing the predator to enter the periplasmic space, between inner and outer membranes. *Bdellovibrio* then reproduces at the expense of host-derived nutrients and eventually lyses the host cell.

Bacillus and *Saccharomyces* represent prokaryotic and eukaryotic genera, respectively, which under unfavorable growth conditions form resistant resting endospores. The sporulation and germination processes are in many respects amazingly similar in the two types of organisms. Members of the *Actinomycetes (Streptomyces, Actinomyces, Nocardia)* form extensive mycelia that so resemble those of fungi that these bacteria were originally classified as fungi until biochemical studies revealed the presence of peptidoglycan-containing cell walls. Members of the *Actinomycetes* also sporulate with the formation of conidiospores in a process not unlike conidiation in *Neurospora.*

While the majority of both bacterial and nucleated cells reproduce asexually by binary fission, yeast and the prokaryote, *Rhodomicrobium vanniellii,* reproduce asexually by budding processes (Figure 1.2). In both species, buds form which lack DNA, and subsequently, after the occurrence of DNA replication in the mother cell, the newly synthesized nucleic acid is transferred to the developing bud. Dual and relatively independent control processes apparently account for the budding and replicative events.

A few bacterial species exhibit spatial as well as temporal control over cellular differentiation. In these species multicellular prokaryotic structures form, and different cells within a population, originally of a single cell type, give rise to different cell types. The type of program to be expressed by any one cell is both temporally and spatially regulated. Examples of such behavior include the blue green bacteria that differentiate into photosynthetic vegetative cells and nonphotosynthetic nitrogen-fixing heterocysts (Figure 1.3). These two cell types arise from the vegetative cell in a carefully regulated sequential fashion, and the mechanism proposed for the spatial regulation of heterocyst development is strikingly similar to that proposed for spatial regulation of development in the coelenterate, *Hydra.* It should be noted that several studies have led to the conclusion that heterocyst development in blue green bacteria is irreversible: a

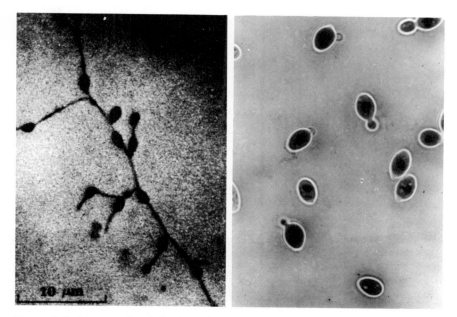

Figure 1.2. Photographs of the bacterium, *Rhodomicrobium vanniellii* (left), and the yeast, *Saccharomyces cerevisiae* (right), during active vegetative growth. Both organisms reproduce asexually by budding. The picture of *R. vanniellii* (left) was taken from E. Duchow and H.C. Douglas, *J. Bact. 58:*411 (1949), copyright American Society for Microbiology, reprinted with permission. The picture of *Saccharomyces cerevisiae* was provided by R. Piñon, Dept. of Biology, University of California, San Diego.

heterocyst cannot divide or revert to a vegetative cell. If this is true, another striking parallel can be drawn between this prokaryotic developmental process and the irreversible differentiation of somatic cells within animal tissues.

One final example of parallel processes occurring in divergent prokaryotic and eukaryotic organisms is the fruiting process in the cellular slime molds such as *Dictyostelium* and in the Myxobacteria. Upon nutrient deprivation, individual cells within these gliding, slime-producing organisms generate and respond to signals that result in massive cellular migration towards an aggregation center. The multicellular aggregates then differentiate with the formation of fruiting bodies containing thousands of resting spores. In both the prokaryote and the eukaryote, fruiting results in the apparently programmed death of those cells destined to form the stalks. Only the spore-forming cells retain the characteristic of immortal potential.

The fact that similar cell patterns and developmental programs are repeatedly observed in the two kingdoms argues in favor of the unifying notion that considerable amounts of genetic material are shared by organisms within the two major biological kingdoms. Additionally, if this assumption is valid, we can expect that many of the molecular mechanisms responsible for present-day de-

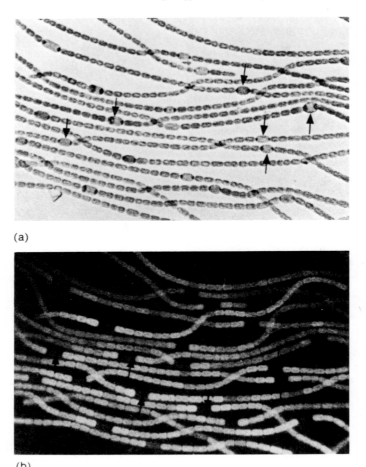

(a)

(b)

Figure 1.3. Photomicrographs of chains of cells of the blue-green bacterium, *Anabaena cyclindrica*. The two cell types depicted are the smaller vegetative cells and the larger heterocysts (indicated by the arrows). While the former cells catalyze photosynthesis, only the latter cells catalyze nitrogen fixation. (a) Transmission image taken with blue light, preferentially absorbed by chlorophyll. (b) Fluorescence image, taken under conditions that specifically reveal the fluorescence of phycocyanine. Heterocysts are barely visible because they lack phycocyanine. Courtesy of M. Donze.

velopmental control must have existed in the simple progenitors of the first organisms to express cyclic or linear programs of differentiation. If this suggestion is in turn valid, we should be able to glean insight into developmental processes in complex organisms by understanding the corresponding processes in the simplest developmental microorganisms. Further, nondifferentiating microorganisms can be expected to possess some of the genetic and biochemical machinery essential to the creation of sequential developmental programs. It should therefore come as no surprise that a detailed understanding of embryogenesis in humans will require that the developmental biologist focus on the simple microorganisms that have provided us with a detailed understanding of metabolic pathways, genetic control mechanisms, and other fundamental processes in molecular biology. In this volume we shall gain a more detailed and universal understanding of development by examining the mechanisms underlying developmental processes in the organisms for which the molecular details are best understood. The relevance of microbial studies to development in higher organisms will be stressed.

Selected References

Balinsky, B.I. *An Introduction to Embryology,* 4th ed., W.B. Saunders Co., Philadelphia, 1975.

Brookbank, J.W. *Developmental Biology: Embryos, Plants, and Regeneration,* Harper & Row, New York, 1978.

Graham, C.F. and P.F. Wareing. *The Developmental Biology of Plants and Animals,* W.B. Saunders Co., Philadelphia, 1976.

Leighton, T. and W.F. Loomis, eds. *The Molecular Genetics of Development,* Academic Press, New York, 1980.

Stanier, R.Y., E.A. Adelberg, and J.L. Ingraham. *The Microbial World,* 4th ed., Prentice-Hall, Inc., Englewood Cliffs, New Jersey, 1976.

Trinkaus, J.P. *Cells into Organs, The Forces that Shape the Embryo,* Prentice-Hall, Inc., Englewood Cliffs, New Jersey, 1969.

Waddington, C.H. *Principles of Development and Differentiation.* Current Concepts in Biology Series, (N.H. Giles, W. Kenworthy, J.G. Torrey, eds.) The Macmillan Co., New York, 1966.

Wessells, N.K. *Tissue Interactions in Development,* An Addison-Wesley Module in Biology, No. 9, Addison-Wesley Publishing Co., Inc., Philippines, 1973.

CHAPTER 2
Molecular Events Accompanying Morphogenesis

The transmigration of life
takes place in one's mind.
What one thinks, that he becomes.
This is the mystery of Eternity.

from the Maitri Upanishad

Morphogenesis, the development of shape, can be discussed both at the organismal level and the cellular level. Within a multicellular embryo several distinct morphogenetic processes occur simultaneously. First, *differential mitosis* results in selective stimulation of the growth of certain cell types relative to others. Selective growth can be due to any one of several cellular or extracellular factors, or to the localization of a cell type within a part of the embryo: (1) differential nutrient supplies or differential abilities of the cell types to transport and accumulate nutrients; (2) the presence of cell-specific hormones and growth factors which stimulate (or inhibit) growth of those cells that possess the requisite receptor proteins; or (3) bioelectric activities, due to the presence of ion selective channel proteins in the plasma membranes of certain cells, which may influence growth rate either by controlling the cytoplasmic ion composition or by creating a transcellular electrical potential. Second, *programmed cell death* plays an important role in morphogenetic processes during embryogenesis. In the developing tadpole, the Rohan-beard cells, which are important constituents of the nervous system, are programmed to die at a certain developmental stage so that other conducting cells can replace them. Loss of the tadpole tail, loss of the webbing between the fingers of the developing human hand, and regression of the Müllerian or Wolfian duct in developing male or female mammals, respectively, represent other examples of programmed cell death allowing embryonic morphogenesis. Third, *cell migration* and *changes in cell shape* occur continuously during development. Cell motility and shape may be controlled by actomyosin-like complexes within the cell, and by microtubules and microfilaments. Both migratory behavior and changes in cell shape may be directed by chemotactic responses or responses to external energy sources. Examples of such behavior during embryogenesis include epiboly (during v ch sheets of cells move inwardly), nerve axon outgrowth, and neural crest migration. Fourth, *intercellular adhesive properties* of cells may change during development. Such associative cell properties give rise to new tissue types and can be growth inhibitory (contact inhibition of growth). Intercellular contacts may prevent migration of the cells (contact inhibition of motion) and may regulate synthesis of new gene products essential to differentiation. Finally, *embryonic induction* and *intercellular modification* represent specialized consequences of cell–cell interactions that are poorly understood at the molecular level but clearly play important roles in embryogenesis.

Virtually all of the processes discussed above which regulate morphogenesis in the developing embryo are demonstrable in microorganisms. Consequently, studies of these processes in well defined, easily controlled microbial systems provide valuable information about morphogenetic processes. Below we shall consider selected examples of microbial morphogenesis that have led to insight into the nature and causes of cellular morphogenesis. In subsequent chapters, the relevant processes mentioned above, which control multicellular morphogenesis, will be discussed in more detail.

Cell-Cycle Dependent Morphogenesis in *Caulobacter*

Numerous developmental processes give rise to polar cells from nonpolar ones. For example, while a fertilized egg is functionally and structurally symmetrical about all axes, embryogenesis gives rise to differentiated epithelial cells with polar characteristics. The plasma membrane on one side of such a cell exhibits a protein composition differing from that on the other side.

Numerous bacteria also exhibit polar characteristics, being asymmetric through all or part of their cell cycles. A few bacteria even generate two distinct asymmetric cell types as part of their normal cell cycle. If certain aspects of the morphogenetic program are interrupted, the cell cycle is inhibited. Thus, morphogenesis and cell division are tightly coupled processes. The most extensively studied prokaryote which displays cell cycle dependent morphogenesis is *Caulobacter crescentus*. Figure 2.1 illustrates the process. Cell shape changes correlate with essential cellular replicative functions. At one end of a mature *Caulobacter* cell is a stalk with an adhesive terminus (the holdfast) that allows the organism to stick to objects or other cells. This adhesive organelle permits *Caulobacter* to remain sessile, and also to form multicellular rosettes that may facilitate sexual conjugation. During the S phase of the cell cycle, during which DNA synthesis occurs, a membrane "whorl" is synthesized at the end of the cell opposite the stalk. This membrane whorl underlies the plasma membrane. Subsequently, during the G2 phase, a single polar flagellum and several short fimbriae are generated at the end of the cell bearing the membrane whorl (Figure 2.2). Only when synthesis of these structures has been completed can cell division occur (M phase), and division gives rise to two morphologically different cell types, a sessile stalked cell, and a motile swarmer cell. The swarmer cell can then swim off to a new location while the stalked cell remains anchored to the substrate. Although the stalked cell is ready to enter S phase after a short G1

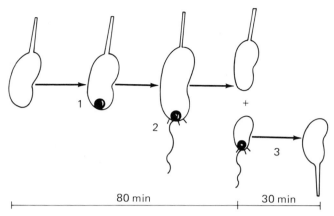

Figure 2.1. Program of cell-cycle dependent morphogenetic events in *Caulobacter crescentus*. From L. Shapiro, N. Agabian-Keshishian, and I. Bendis, *Science 173*:884–892 (1971); copyright 1971 by the AAAS, reprinted with permission.

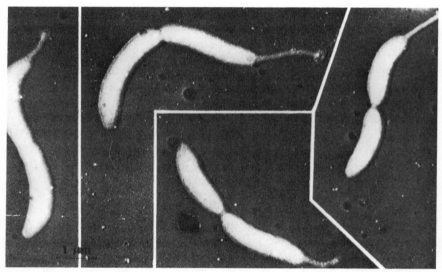

Figure 2.2. Photograph of bipolar cells of *Caulobacter crescentus* just prior to cell division. From A.E. Houwink, *Antonie van Leeuwenhoek 21*:54 (1955), copyright Netherlands Society for Microbiology, reprinted with permission.

phase, the swarmer cell must undergo extensive changes in morphology: The membrane whorl and fimbriae disappear; the flagellum falls off intact and can be recovered in the culture fluid; and a stalk is generated at the site formerly occupied by the flagellum. This last process is accompanied by an increase in the rate of cell wall synthesis. All of these events occur during the G1 phase of the cell cycle and represent prerequisites for chromosomal replication. Thus, while the stalked daughter cell is characterized by a minimal division time of 80 min, the swarmer cell requires at least 110 min. to complete its cell cycle because it must remain in the G1 phase throughout the swarmer cell-to-stalked cell transition.

Several features of *Caulobacter* render it excellent material for a detailed study of development: First, unlike *E. coli* and other genetically well-characterized bacteria, the *Caulobacter* cell cycle contains visible morphogenetic events. Second, the generation time of the organism is short (80 min for the stalked cell; 110 min for the swarmer cell at 30°C) facilitating the preparation of cellular material for biochemical analyses. Third, *Caulobacter* can be grown in a defined medium, so that conditions influencing morphogenesis can be controlled rigorously. Fourth, mutants can be isolated easily by the standard techniques of the bacterial geneticist, and both conjugational and transductional analyses of the mutations are possible. Finally, *Caulobacter* can be grown in synchronous cultures since stationary phase cells are always arrested as stalked cells in the G1 phase of the cell cycle. Transfer of the arrested cell population to fresh medium allows entry into the normal cycle with a 10 min lag phase so that cell division occurs in 90 min with the generation of one swarmer and one stalked cell per parent. It should be noted that while these events are cyclic, irreversible steps prevent reversal of the pathway. Further, biochemical analyses have revealed that certain gene products are synthesized only during select periods in the cell cycle while others are synthesized continuously. As for morphogenesis during embryonic development, both reversibly and irreversibly controlled events affecting gene expression determine the obligatory morphogenetic sequence that is an integral part of the cell cycle in *Caulobacter*.

Examination of events accompanying and possibly controlling the cell cycle has revealed that cyclic nucleotides can regulate growth and the synthesis of certain cellular structures in *Caulobacter*. The cytoplasmic ratio of cyclic AMP to cyclic GMP may be of particular importance. Separate cyclic AMP and cyclic GMP binding proteins have been identified, and their involvement in regulation has been suggested. Another interesting observation deals with synthesis of the flagellar filament. This structure consists of polymerized flagellin monomers; each flagellum consists of two flagellin monomers; one predominates proximal to the cell and the other predominates distal to the cell body. It appears that a cell-cycle dependent genetic switch occurs, which shuts off synthesis of one flagellin and turns on synthesis of the other. Such switch mechanisms may be characteristic, but not exclusive of developmental systems.

Control of Vegetative Morphogenesis in *Neurospora*

Neurospora crassa is a genetically well-characterized fungus that can generate an extensive mycelial mat from a single haploid spore. Spore germination results in the formation of a germ tube (Figure 2.3), which extends to become the mycelial hyphae of the growing colony. DNA replication is followed by incomplete septum formation, yielding cross walls through which nuclei can pass from

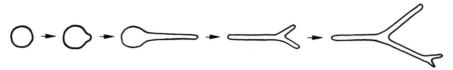

Figure 2.3. Schematic depiction of conidiospore germination in *Neurospora crassa*.

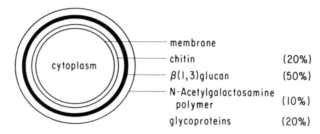

Figure 2.4. Cross-sectional depiction of the multilayered structure of the cell wall of the vegetative *N. crassa* mycelium.

cell compartment to cell compartment as a result of cytoplasmic streaming. Branch points occur fairly regularly along the hyphae. Metabolism is most active at the tip where growth occurs, and branching is always apical.

Cross-sectional examination of *Neurospora* hyphae reveals a multilayered cell wall (Figure 2.4). Adjacent to the plasma membrane is a thin chitin layer (20% of the cell wall by weight) in which the repeating unit is an N-acetylglucosamine residue. Biosynthesis of chitin requires synthesis of the "activated" N-acetyl-glucosamine precursor, UDP-N-acetylglucosamine. External to the chitin layer is a thick glucan layer (50% of the wall dry weight) in which the glucose residues are polymerized in β-1,3 linkages. UDP-glucose is the activated precursor of this polymer. Finally, an external layer consisting of an N-acetylgalactosamine polymer (10%) and a number of glycoproteins (20%) confers upon the wall its antigenic and species-specific characteristics. Each of these layers is important to the integrity of the wall.

A large number of morphological (*morph*) mutants have been isolated and characterized in *Neurospora crassa*. These mutants exhibit normal growth rates but abnormal colony morphology. Auxotrophs, with reduced growth rates, typically show normal colony morphology. It is, therefore, clear that genes controlling mycelial morphology are distinct from those controlling growth. All *morph* mutants so far characterized have been shown to be the result of recessive point mutations with stable, non-auxotrophic phenotypes. The mutations have been mapped to seven distinct linkage groups within the *Neurospora* genome.

Morph mutations generally alter mycelial morphology in a uniform way: First, the apical growth rate slows, and the diameter of the hyphae increases in a compensatory fashion such that the differential increase in cell mass per unit time is not altered. These changes result in a more bulbous hyphal compartment. Second, the branching frequency (the number of branches per unit length) increases, giving rise to an increased density of growth. The angle of branching remains unchanged. The mycelial morphologies of a number of different *morph* mutants are illustrated in Figure 2.5.

Morph mutants have been characterized biochemically and have been found to be deficient, but not completely lacking, in the activities of enzymes in three

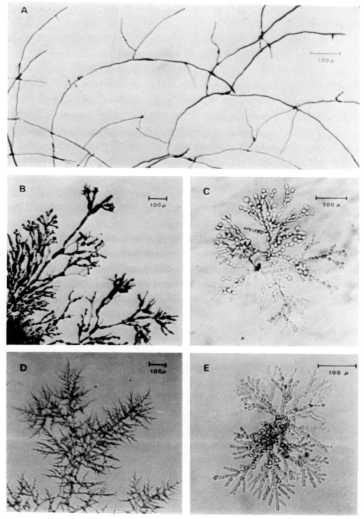

Figure 2.5. Morphologies of the mycelia of wild type and *morph* mutants of *N. crassa*. Cells were grown on agar medium at 24°C. (A) Edge of a wild type colony; (B) *Frost* mutant (glucose 6-phosphate dehydrogenase); (C) *Colonial-2* mutant (glucose 6-phosphate dehydrogenase); (D) *Balloon* mutant (glucose 6-phosphate dehydrogenase); (E) *Ragged* mutant (phosphoglucomutase). From S. Brody in *Developmental Regulation: Aspects of Cell Differentiation* (S. Coward, ed.), copyright 1973, Academic Press, reprinted with permission.

major classes (Table 2.1). Class I mutants are defective for phospholipid bio-synthetic processes. One of these mutants cannot methylate phosphatidylethanol-amine efficiently, and therefore makes less than normal amounts of phosphatidylcholine. The other is defective for inositol phosphate synthetase, and consequently synthesizes decreased amounts of phosphatidylinositol. These two types of mutants may show alterations in their cell wall compositions, but the defects can be completely reversed by growth in the presence of choline or inositol, respectively.

Class II mutants show decreased activities for either glucose 6-phosphate dehydrogenase or 6-phosphogluconate dehydrogenase. As a result, they reduce $NADP^+$ at decreased rates, and the substrates of these two enzymes, glucose 6-phosphate or 6-phosphogluconate, accumulate. Probably because of the decreased availability of reducing power, the membrane content of the principal polyun-saturated fatty acid, linolenic acid, is diminished, and possibly because of the resultant decreased fluidity of the membrane, cell wall precursors may be as-sembled abnormally. One of these mutants (colonial-2) is temperature sensitive, and a 10°C temperature shift, resulting in partial loss of glucose 6-phosphate dehydrogenase activity, gives rise to a rapid increase in branching frequency.

The third class of mutants (Class III in Table 2.1) consists of mutants with defects in three enzymes that function in the biosynthesis of cell wall precursors. Phosphoglucomutase interconverts glucose 6-P and glucose 1-P, the metabolic precursor of UDP-glucose, and UDP-glucose is the biosynthetic precursor of cell wall glucan. Phosphoglucoisomerase interconverts glucose 6-P and fructose 6-P. The latter compound is a required substrate for the synthesis of UDP-N-acetylglucosamine, the biosynthetic precursor of chitin. And UDP-N-acetylglu-cosamine epimerase catalyzes the interconversion of UDP-N-acetylglucosamine

Table 2.1. Biochemical Characteristics of *morph* mutants of *Neurospora crassa*

Class	Gene designation	Primary defect	Deficiency in membrane or wall
Ia	choline	Phosphatidyl ethanolamine —methyltransferase	Phosphatidylcholine content decreases; phosphatidylethanolamine content increases
Ib	Inositol	Inositol phosphate synthetase	Phosphatidyl inositol content decreases; chitin content decreases.
IIa	Colonial-2, balloon, frost	Glucose 6-P dehydrogenase	Linolenic acid content decreases; altered cell wall
IIb	Colonial-3 Colonial-10	6-phosphogluconate dehydrogenase	Linolenic acid content decreases; altered cell wall
IIIa	Ragged	phosphoglucomutase	Glucan content decreases
IIIb	—	phosphoglucose isomerase	Chitin content decreases
IIIc	Doily	UDP N-acetyl glucosamine epimerase	N-Acetylglucosamine polymer content decreases

and UDP-N-acetylgalactosamine which is required for the synthesis of the cell wall N-acetylgalactosamine polymer (Figure 2.4).

Thus, all three classes of *morph* mutants directly affect the compositions of the hyphal membrane or cell wall, and those which primarily affect cell membrane composition secondarily alter the cell wall composition. Deviation from the normal composition of the cell wall might be expected to decrease the structural rigidity of the wall, thereby diminishing resistance to internal pressure. If it is assumed that a general weakening of the wall would give rise to a less narrow (more bulbous) hyphal filament in response to internal pressure, and that a local weakening of the wall at the growing tip would permit more frequent branching of the mycelium, we can account for the morphological changes resulting from defects in the cell wall biosynthetic machinery. These types of studies therefore allow us to identify and understand the mechanisms controlling and altering mycelial morphology in *Neurospora*.

Ion Currents Associated with Morphogenesis in Brown Algae

Developmental biologists have long suspected that electric fields and/or ion currents may influence growth and development by "polarizing" the cytoplasm of a single cell or groups of cells. Only fairly recently, however, have sufficiently sensitive electrophysiological and biochemical techniques been applied to developing systems to verify these hypothetical relationships. One of the most thoroughly studied of such systems from this standpoint is the development of eggs of the marine brown algae, *Fucus* and *Pelvetia*. These organisms live in the intertidal region and can reach lengths of several meters. However, the fertilized egg cells of these algae are only about 0.1 mm in diameter, and shortly after fertilization they appear as totally undifferentiated spheres (Figure 2.6a). About 12 hr after fertilization of the egg cell by the sperm, germination occurs: a distinct protrusion appears at one side of the cell, and it elongates over the next several hours (Figure 2.6, b–d). During this process, the zygote divides into two morphologically and functionally distinct cells, one called the *rhizoid* cell, which includes the original protrusion and will differentiate into a *holdfast* that anchors the organism to the substratum, the other called the *thallus* cell, which will develop into the remainder of the organism.

A number of factors can influence on which side of the fertilized egg of *Pelvetia* or *Fucus* germination will occur. These factors include temperature, light, electric currents, and pH gradients. For example, if light is shone on one side of the zygote, the rhizoid almost invariably develops on the dark side of the embryo. This process is reversible up to about 10 hr after fertilization. If the direction of illumination is reversed within this period, germination will occur on the opposite side of the second light beam. Ten hours after fertilization, however, the embryo becomes "committed" to a specific polarity of development even though no distinct morphological features are visible in the zygote at this time, and germination does not occur for an additional 12 hr. If the direction of

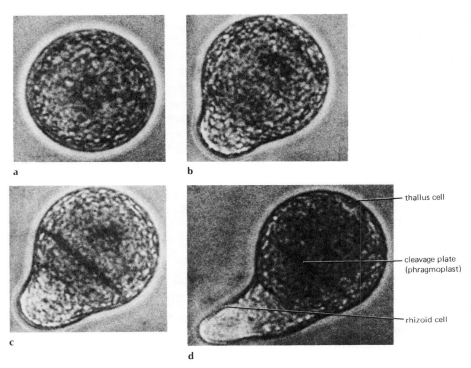

a

b

c

d

thallus cell

cleavage plate
(phragmoplast)

rhizoid cell

Figure 2.6. Stages in the development of a fertilized *Fucus* egg. The micrographs show the cell at 4hr(a), 16hr(b), 18hr(c) and 26hr(d) after fertilization. In (a), the cell appears as a totally undifferentiated sphere, while in (b)–(d), various stages in the process of germination are shown. In (d), the zygote has divided into a thallus and a rhizoid cell. Courtesy of G.B. Bouck.

light is reversed during this period (10–12 hr postfertilization), the rhizoid cell will still develop on the side opposite to that of the first light beam.

These observations suggest that between 0 and 10 hr postfertilization, a reversible process takes place which can be influenced by environmental factors and which determines the eventual site of germination. A likely candidate for this process was discovered in *Fucus* eggs in 1966. Developing embryos that were aligned in one direction in a capillary tube by a beam of light were able to drive an electric current through themselves and the capillary as shown in Figure 2.7. More recent measurements have used a small vibrating platinum electrode, which can be placed near a single zygote as shown in Figure 2.8A. This instrument compares the potential at one end of its oscillation to that at the other by means of a built-in reference electrode. By positioning this probe at different places around the developing *Pelvetia* embryo, it has been possible to measure directly the electric fields surrounding the fertilized algal egg at different stages of its development.

By means of such experiments, it has been shown that an inwardly directed

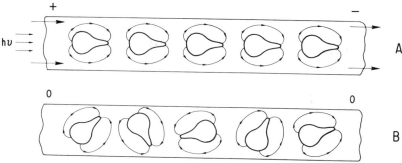

Figure 2.7. Measurement of a current driven through developing *Fucus* eggs. Cells were aligned in a capillary tube by a beam of light (A). A current was shown to be driven through the capillary under these conditions (straight arrows). In contrast, if cells were not aligned by light, no current was driven through the tube (B). Current patterns around the cells that were inferred from these experiments are shown by the curved arrows. (Adapted from L. Jaffe, *Proc. Natl. Acad. Sci. USA 56:*1102–1109 (1966).)

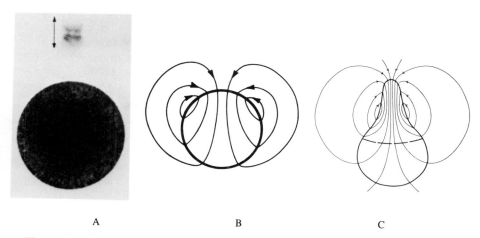

Figure 2.8. Use of a vibrating platinum electrode to measure current patterns around a developing *Pelvetia* egg. (A) Micrograph of the vibrating electrode (arrow shows the axis of vibration) near a *Pelvetia* egg 3 hr after fertilization. (B) Steady current pattern inferred by making measurements with the probe at various positions around the egg cell. This pattern was determined in a cell just before germination, but an inwardly directed current of positive charge at the presumptive rhizoid end can be detected as early as 30 min after fertilization. (C) Current pattern around a germinated *Pelvetia* egg at the two-celled stage. This current appears in pulses rather than in the more steady manner seen in cells before germination. A and B were from R. Nuccitelli, *Devel. Biol. 62:*13–33 (1978), copyright Academic Press, reprinted with permission; C was reproduced from *The Journal of Cell Biology,* 1975, Vol. 64, pp. 636–643, by copyright permission of The Rockefeller University Press.

current of positive charge predicts the point of germination in *Pelvetia* as early as 30 min after fertilization (Figure 2.8B) even though germination takes place much later. Furthermore, current pulses of similar polarity (i.e., with positive charges entering the growing tip) take place in *Pelvetia* embryos even after germination in the 2-cell stage (Figure 2.8C). Electrophysiological experiments have shown that much of the early current detectable in undifferentiated zygotes is due to a current of Ca^{2+} ions entering the egg membrane in an area in which germination will eventually occur (rhizoid end) and leaving through most of the rest of the cell surface (thallus end). Furthermore, if the polarity of germination is reversed, for example by shifting the direction of illumination of the embryo, corresponding changes in the current pattern also occur as long as this is done within 10 hr after fertilization. These observations strongly suggest that a current of Ca^{2+} may be directly involved in determining the polarity of differentiation in a *Pelvetia* zygote.

Ionic currents have also been detected in a number of other developmental systems. For example, germinating pollen grains, developing animal zygotes, and even regenerating amphibian limbs have associated ion fluxes that accurately predict the polarity of the differentiation process. In addition to Ca^{2+}, trans-membrane movements of Na^+, K^+, and Cl^- appear to be important components of these currents depending on the specific system studied. An important question arising from these observations is whether these currents are a primary cause of a particular developmental process or whether they are a consequence of these events. In *Pelvetia,* the appearance of such currents well before any morphological changes can be detected argues in favor of their causal role. Furthermore, the polarity of developmental processes can often be influenced by imposed electric fields. Imposition of a current across a fertilized *Pelvetia* egg usually results in a site of germination predicted by the polarity of the ionic currents shown in Figure 2.8B. Even more remarkably, the regeneration of amputated limbs of amphibians is accompanied by an outward flow of positive charges from the limb stump, and steady currents imposed through the affected area have been shown to influence the rate and extent of regeneration! Thus, it seems likely that ion currents are at least one cause of subsequent developmental changes and not simply byproducts of them.

The Generation and Roles of Ion Currents in Development

In order for currents of the types illustrated in Figure 2.8 to be generated, functionally active ion channels and pumps must be asymmetrically distributed in the cell membrane. If most of the early current detectable in *Pelvetia* eggs (Figure 2.8B) is due to Ca^{2+} fluxes, it is probable that active Ca^{2+} channels are localized near the prospective rhizoid end, while active Ca^{2+} pumps generate outward Ca^{2+} fluxes at other areas of the cell membrane. External stimuli such as light must, therefore, be able to influence the distribution of these pumps and channels, although the mechanisms by which they do this are unknown. Similarly, transcellular currents found in other developing biological systems must either rely on the localization of ion transport proteins in specific regions of the cell membrane or on selective activation of these channels in specific parts of the cell.

If we accept the proposal that ion currents play a role in the development of an initially undifferentiated cell or group of cells, then a reasonable question to pose is: How do such currents act to polarize the cytoplasm, leading to the differentiation of different types of cells? Returning to the example of *Pelvetia,* several possible answers to this question can be entertained. For example, a transcellular current of Ca^{2+} ions of the type shown in Figure 2.8 should lead to an *intracellular Ca^{2+} gradient,* with the concentration of this ion being highest near the prospective germination site. Since many cellular processes are highly sensitive to the concentration of Ca^{2+} (contractile systems, membrane vesicle fusion, etc.), their activities within the cell could be spatially oriented by a gradient of this ion. Furthermore, a *transcytoplasmic electric field* can also be the result of such current fluxes. In *Pelvetia,* secretory vesicles involved in the deposition of new cell wall and membrane material move to the site of future germination within 6 hours after fertilization of the egg, during the time when a steady transcellular current is detectable. It is, therefore, conceivable that negatively charged vesicles could be "electrophoresed" to the rhizoid end of the cell as a consequence of an electric field across the cytoplasm of the single-celled zygote.

A dramatic example of the latter type of mechanism has been found in the *öocytes* (immature developing egg cells) of the moth, *Hyalophora cecropia.* In the female insect, these cells are surrounded during early stages of their development by seven *nurse cells* derived, as is the öocyte, from divisions of the primary germ cell, the *öogonium.* These nurse cells are connected to the öocyte by cytoplasmic bridges that result from incomplete cell divisions of the öogonium (Figure 2.9). Electrical potential measurements have established that the öocyte cytoplasm is approximately 10 mV more positive than that of the nurse cells. If this is the case, then negatively charged molecules should tend to flow into the öocyte from the nurse cell cytoplasm while the opposite should be true for positively charged molecules. This prediction was confirmed by injecting öocyte–nurse cell complexes with fluorescently labeled acidic and basic proteins, serum globulin and lysozyme, respectively. Serum globulin was shown to be electrophoresed from nurse cells to the öocyte, while lysozyme remained in the nurse cell cytoplasm (Figure 2.9).

One of the major functions of the nurse cells in insect öocyte development is to contribute to the rapid growth rate of the egg cell by synthesizing large amounts of macromolecules such as ribosomal RNA. Apparently, a potential is maintained across the cytoplasmic bridges in part to facilitate the movement of such negatively charged molecules into the öocyte cytoplasm. As is the case in the fertilized *Pelvetia* egg cell, this electric field is most likely the consequence of transcellular ionic currents generated around, and through, the öocyte–nurse cell complex by localized ion transport systems.

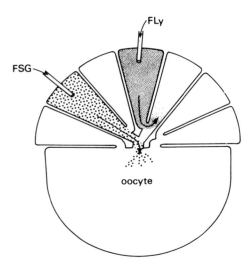

Figure 2.9. Diagram of the *Hyalophora cecropia* öocyte–nurse cell complex. Seven nurse cells are connected to the larger öocyte by cytoplasmic bridges. Fluorescently labeled serum globulin (FSG), an acidic protein, is electrophoresed into the öocyte after injection into a nurse cell. In contrast, fluorescent lysozyme (FLy), a basic protein, is observed to remain within the nurse cells after injection. This is explained by the fact that the öocyte cytoplasm is about 10 mV more positive than that of the nurse cells. Reprinted by permission from *Nature,* Vol. 286, pp. 84–86. Copyright © 1980 Macmillan Journals Limited.

Although we have limited our discussion of the role of ion currents in development to these two well-studied examples, it is probable that such currents as well as more transient electric fields associated with changes in membrane potentials will have general importance in influencing cellular differentiation and growth in the biological world. For example, transient depolarization of the egg cell membrane following fertilization has been observed in a number of systems. This phenomenon appears to be involved in preventing fertilization by a second sperm cell (*polyspermy*), and ionic currents associated with it may have a role in initiating development of the zygote in some organisms (see Chapter 10). It is also possible that germination of spores and seeds, germ tube formation in *Neurospora* and "Shmoo" morphogenesis from vegetative yeast cells, induced by yeast pheromones in preparation for zygote formation (Chapters 9 and 10), are directed by transcellular ionic currents. Examination of these possibilities remains for future investigations.

Selected References

Ashworth, J.M. and J.E. Smith, eds. *Microbial Differentiation,* Society for General Microbiology Symposium 23, Cambridge University Press, 1973.

Brody, S. "Genetic and Biochemical Studies on *Neurospora* Conidia Germination and

Formation" in *The Fungal Spore: Morphogenetic Controls* (G. Turian and H.R. Hohl, eds.), Academic Press, London, 1981.

Harold, F.M. Ion currents and physiological functions in microorganisms, *Ann. Rev. Micro. 31:*181 (1977).

Jaffe, L.F. and R. Nuccitelli. Electrical controls of development, *Ann. Rev. Biophys. Bioeng. 6:*445 (1977).

Leighton, T. and W.F. Loomis, eds. *The Molecular Genetics of Development,* Academic Press, New York, 1980.

Robinson, K.R. and L.F. Jaffe. Polarizing fucoid eggs drive a calcium current through themselves, *Science 187:*70 (1975).

Schmit, J.C. and S. Brody. Biochemical genetics of *Neurospora crassa* conidial germination, *Bact. Rev. 40:*1–41 (1976).

Shapiro, L. Differentiation in the *Caulobacter* cell cycle, *Ann. Rev. Micro. 30:*377 (1976).

Shapiro, L., P. Nisen, and B. Ely. "Genetic Analysis of the Differentiating Bacterium: *Caulobacter crescentus*" in *Genetics as a Tool in Microbiology,* Society for General Microbiology Symposium 31 (S.W. Glover and D.A. Hopgood, eds.), Cambridge University Press, 1981.

CHAPTER 3

Cellular Mortality, Growth Regulation, and the Phenomenon of Cancerous Transformation

What a piece of work is man!
How noble in reason!
How infinite in faculty!
In form and moving how express and admirable!
The paragon of animals!

Shakespeare

The Lord looks down from heaven
on all mankind
to see if any act wisely,
if any seek out God.
But all are disloyal, all are rotten to the core;
not one does anything good,
no, not even one.

Psalms 14

Over the past few decades a huge research effort has been devoted to understanding the mechanisms of cancerous transformation. This effort has led to recognition of the fact that the unrestricted growth of most cancer cells is caused by release of normal cells within the body from the regulatory constraints imposed upon them. An understanding of cancerous transformation therefore leads to knowledge of the events controlling normal cell proliferation. In this chapter we shall first discuss the possible evolutionary origins of differentiated animal cells, both somatic (tissue) cells and germ (reproductive) cells. Subsequently, we shall consider the events that occur when normal cells become cancerous. It will become apparent that two distinct events frequently (but not always) accompany the transformation process: loss of sensitivity to *growth regulation*, and loss of the constraints of *cellular mortality*. Further, these two events are clearly distinct from each other, as well as from the genetic events that determine expression of the differentiated state. Although the phenomenon of cell mortality sometimes referred to as *programmed cell death* is presently poorly understood, it seems to have evolved as the terminal step in a sequence of irreversible differentiation events.

Close relationships between growth regulation, cancerous transformation and expression of the differentiated traits of somatic cells appear well established.

Moreover, since multicellular organisms possess both somatic (mortal) and germ (immortal) cells, it may not appear surprising that two fundamentally different types of events may give rise to cancerous transformation, depending on the cell of origin. These relationships and the mechanisms that may be involved are discussed in the subsequent sections of this chapter.

Evolution of Multicellularity

The first biological cells to exist on earth some 3 billion years ago were probably very simple, both structurally and functionally. They may have been simplified forms of present-day anaerobic bacteria. These cells merely reproduced their kind by binary fission and were capable of indefinite growth under favorable environmental conditions. No genetic programs involving irreversible decision making gave rise to temporal changes in cell structure or function. Programmed mortality had not yet been invented, and no regulatory constraints prevented rapid cell proliferation under favorable conditions.

The first differentiating cells were probably those that carried out cyclic mechanisms for the generation of resting cells, haploid spores that were less sensitive to extremes of environmental conditions than were the vegetative cells (Chapter 4). Very probably, pre-existing and novel mechanisms of genetic rearrangement (Chapters 11–13) and control (Chapters 4 and 7) were utilized in order to achieve a dormant cell type. Entry into the sporulation program had to be triggered by some adverse external condition. Sporulation type mechanisms evolved into three major categories: endosporulation, where the entire process occurred within a vegetative cell that itself remained submerged in an aqueous solution; aerial sporulation, which involved temporal and spatial differentiation of the cell population into spores and mycelial structures bearing the spores above the aqueous surface; and fruiting body formation, which provided the spores with a protective shell and facilitated delayed dissemination. Examples of these three types of sporulation processes are found in representatives of the prokaryotic world: *Bacillus, Streptomyces,* and *Myxococcus,* respectively. Because of the increasing degree of complexity of these three types of sporulation processes, we may speculate that they probably evolved in the order described. The last of these processes, resulting in the evolution of extensive fruiting structures, may have been the first example of primitive multicellularity.

Anaerobic photosynthetic bacteria probably provided the first mechanism for utilizing solar energy. Possibly a single pigment protein, such as bacteriorhodopsin, found today in certain halophilic bacteria, was initially involved, but later a more complex mechanism appeared involving photon capture, electron activation, electron transfer, and the evolution of oxygen gas. Some environments were poor in sources of reduced nitrogen, and the need for such compounds for growth conferred an evolutionary benefit on those organisms that could also reduce molecular nitrogen. However, nitrogenase, the enzyme responsible for nitrogen fixation, could function only in an anaerobic or microaerophilic environment because oxygen gas strongly inhibited its activity. In order to solve this problem, photosynthetic cyanobacteria, blue-green bacteria, evolved a mechanism for the development of a unique cellular compartment for nitrogen fixation that was separated from, but connected to the photosynthetic, oxygen-evolving compartments. Deprivation of reduced nitrogen triggered a differentiation sequence which, instead of giving rise to a thick-walled resting spore, gave rise to a thick-walled, nitrogen-fixing heterocyst, poorly permeable to oxygen. Because some of the cells in the population remained functional for photosynthesis and vegetative growth, the formation of heterocysts represented a new evolutionary form of spatial differentiation wherein one population of cells of uniform structure and function gave rise to two distinct cell types. While heterocyst formation in the blue-green bacteria may represent one of the more complex forms of prokaryotic differentiation, its significance can only be appreciated if one considers that it provided the evolving eukaryotic kingdom with genetic mechanisms for initiating much more complex sequences of spatial differentiation. Heterocyst formation may also have been one of the first examples of an irreversible sequence of genetic events in which a nondividing, and thus mortal cell was the end product of differentiation.

The appearance of eukaryotes by virtue of degenerative co-evolution of endosymbiotic organisms and their hosts may have occurred repeatedly in nature (parallel evolution), or the different eukaryotic cell types may have arisen from a single lineage (linear, sequential evolution). Extensive genetic exchange between species may have occurred during this period giving rise to eukaryotes with all of the adaptive mechanisms found in the differentiative prokaryotes. Thus, endosporulation (*Saccharomyces*), conidiation (*Neurospora*), and fruiting (*Dictyostelium*) appeared in eukaryotes as adaptive mechanisms that allowed the different species to withstand extremes of environmental conditions. Multicellularity, already advantageous to cyanobacteria and myxobacteria, provided eukaryotes with much greater protection from environmental threats because it allowed the developing organism to create its own more favorable extracellular environment, thereby facilitating the maintenance of intracellular homeostasis favoring continued growth.

The evolving haploid eukaryotic cells probably lacked a cell wall and possessed fairly fluid membranes, which in part allowed occasional fusion of cells with the formation of new, larger cells with double the haploid number of chromosomes. The diploid state was immediately advantageous because it circumvented the deleterious effects of recessive mutations, and it was of evolutionary benefit because the availability of two copies of each gene allowed loss of function on one chromosome with retention of function on the other. By this means, each diverging gene gained the potential for acquisition of new functions without organismal loss of the original function. Hence, evolutionary advances were greatly accelerated.

In order to facilitate genetic exchange and recombination between potentially advantageous mutations, the development of meiosis, genetic reduction to the haploid state, was necessitated. Subsequently, fusion of two haploid cells could restore the diploid state, but with new combinations of the mutated haploid genomes. The diploid (or polyploid) state therefore evolved for the vegetative cell, giving rise to multicellularity, while the haploid, less complex cell retained the essential reproductive functions. The advent of two cell types (somatic and germ cells), both of which might be derived from an undifferentiated stem cell, allowed the profusion of irreversible differentiative events within the somatic cell population. Irreversible differentiation gave rise to cell mortality within the former group of cells, which was permissible only because the essential function of species propagation was maintained by the germ cells. The notion of organismal differentiation and hence organismal mortality resulted from identification of the organism with the somatic cell population. Thus, each evolving organism came to possess both mortal and immortal cells within a single multicellular structure, and the organism was said to die when the somatic cells within the organism ceased to live.

It should be noted that unicellular organisms, such as the ciliates, evolved an alternative mechanism for allowing irreversible differentiation with retention of germ cell function, all maintained within a single *multinucleate* cell. The irreversible differentiative events were recorded and controlled in the macronucleus while the micronucleus was retained for reproductive reasons (see Chapter 13). In both the unicellular ciliates and the evolutionarily more successful multicellular organisms, sexual conjugation serves as the rejuvenating principle.

Two Fundamental Types of Cancerous Transformation

The progression of evolutionary events discussed above can be summarized as follows: Immortal cells gave rise to differentiating cells which were programmed for, or committed to, a linear sequence of irreversible steps in development. The terminal developmental state for any such cell is death. This same phylogenetic progression is applicable to ontogenetic development in a multicellular animal (see Figure 3.1). A fertilized egg, which is immortal, gives rise to tissue cells programmed to die as well as to germ cells, which retain the potential to live forever.

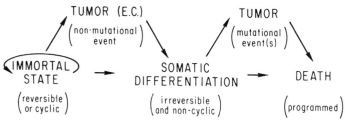

Figure 3.1. Proposed scheme for the evolution of cyclic and irreversible developmental processes during phylogeny and ontogeny. Irreversible differentiation is believed to have given rise to programmed cell death.

Since two fundamentally different cell types exist in the body, it is not surprising that two distinct mechanisms can apparently give rise to unregulated growth. In the first transformation mechanism, *reversible* malignancy results from tissue disorganization of totipotent stem cells (germ cells or early embryos), and unregulated growth of these undifferentiated cells ensues. No mutational event need occur, so the genetic material of the cell is unaltered. In the second transformation mechanism, *irreversible* malignancy results from genetic alteration (mutation) of partially or fully differentiated somatic cells. This type of mechanism is considered to be irreversible because the cell is genetically altered by the transformation process, and restoration of normal environmental conditions will not restore sensitivity to growth regulation. In the next two sections we shall consider these two transformation processes separately.

Teratocarcinogenesis—Reversible Malignancy

A scheme for the process of *teratocarcinogenesis,* in which a germ cell can give rise to a malignant cancer, is shown in Figure 3.2. The cell of origin can be a primordial germ cell or a young embryo. In both cases, tissue disorganization rather than a mutational event is thought to give rise to malignancy. The primordial germ cells are the diploid spermatogonia, present in the testes of a male, which after meiosis give rise to sperm cells, and the öogonia, present in the ovaries of a female, which after meiosis give rise to the haploid egg cells.

There are several independent lines of evidence bearing on the cellular origin of the teratoma. First, teratomas occur most commonly in the gonads, but occasionally arise elsewhere in the body, probably by metastasis (migration through the blood stream). Second, testicular teratomas can be induced experimentally in male birds, but only in the spring when the spermatogonia divide. Third,

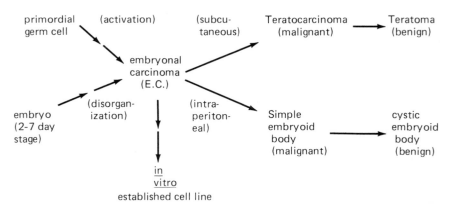

Figure 3.2. Proposed scheme for teratocarcinogenesis as it is believed to occur in mice and other multicellular organisms.

almost all diploid testicular teratomas have the XY sex chromosomes while most ovarian teratomas are XX. Finally, grafting 2- to 64-celled ova to certain abnormal body locations can give rise to a teratocarcinoma, a tumor which is said to be pleuripotent because it possesses the potential to generate many differentiated cell types. Transformation of adult somatic cells never gives rise to a teratocarcinoma.

The term given to the pleuripotent and malignant stem cell is *embryonal carcinoma*. If a population of such cells is injected subcutaneously into an animal, a teratocarcinoma will develop with actively growing, nondifferentiated stem cells as well as benign, fully differentiated and nondividing tissue cells. All of the cells in the tumor arose from the embryonal carcinoma cell, which by virtue of its pleuripotency was able to give rise to many tissue types. The eventual destiny of a *teratocarcinoma* is complete differentiation to a *teratoma*. A teratoma is benign because it lacks the pleuripotent stem cells and is therefore in a state of growth stasis.

Figure 3.3 shows a schematic depiction of a teratocarcinoma while Figure 3.4 shows photographs of the different tissue types from such a tumor. Within a teratocarcinoma can often be found ectoderm (skin and neural tissue), mesoderm (muscle, bone, cartilage), endoderm (gut), and extraembryonal tissue (trophoblasts, yolk sac cells). The small embryonal carcinoma stem cells are also present. Interestingly, tissue types found together in the body are frequently in association in the tumor. Thus, one might find intestinal epithelia surrounded by fibroblasts and collagen fibers with the latter in association with well differentiated, multinucleate muscle fibers. Cartilage, bone marrow, and bone are frequently ob-

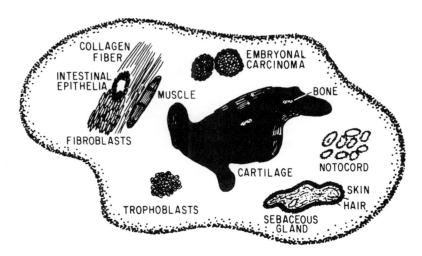

Figure 3.3. Schematic depiction of a teratocarcinoma with a variety of differentiated cell types as well as undifferentiated stem cells. Note the clustering of cell types normally found together in the body.

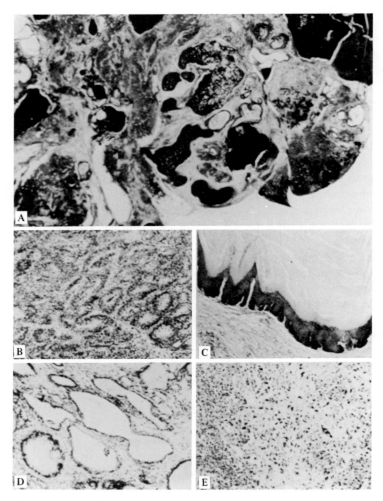

Figure 3.4. Photographs of portions of a single teratoma. A variety of tissue types is shown after tissue sections were stained with hematoxylin, eosin, and alcian blue. (A) A variety of differentiated tissues is apparent at low magnification. Cartilage (mesodermal derivative) is readily recognizable as the most darkly stained tissue. Magnification: $\times 17$. (B) Early neural differentiation (ectodermal derivative). Magnification: $\times 100$. (C) Keratinizing epithelium (skin: an ectodermal derivative). Magnification: approximately $\times 50$. (D) Endodermal cysts. Magnification: $\times 130$. (E) Trophoblast. Extremely large cells of extra-embryonic origin. Contrast with typical embryonal carcinoma cells (the stem cells of the tumor) shown on the left and lower right. Magnification: $\times 110$. From G.R. Martin, *Cell 5*:229–243 (1975), copyright M.I.T., reprinted with permission.

served together as are keratinized epithelia (skin and hair) with subaceous glands. This observation suggests that one tissue type within the tumor can influence the differentiation process in nearby cells.

Injection of embryonal carcinoma cells into the peritoneal cavity of an animal gives rise to malignant *"simple" embryoid bodies* which fully differentiate into benign *"cystic" embryoid bodies* (Figure 3.2). This process is similar to that which occurs after subcutaneous injection of the cancerous stem cells. From this fact it appears that teratoma differentiation is not dependent on the site of injection.

Cell lines that can be maintained in tissue culture indefinitely have been derived from embryonal carcinoma cells. These cell lines either retain a particular tissue-specific differentiated character or they retain some degree of pleuripotency so that the stem cell can differentiate into more than a single tissue type. Studies with these cells in the tissue culture environment have established that expression of the programs of differentiation does not depend on the environment of the host tissue and therefore must be determined by the genetic composition of the cell. But local influences are also important. The process of differentiation appears to occur irreversibly and results in loss of malignancy.

Recent experiments have provided evidence for the conclusion that some embryonal carcinoma cells are genetically normal, and that teratocarcinogenesis therefore has a nonmutational basis. This conclusion was based upon the demonstration that a formerly malignant mouse embryonal carcinoma cell could participate in embryogenesis and retained the ability to express all of the normal programs of differentiation. The following experimental protocol was employed: a particular line of embryonal carcinoma cells which had been maintained in continuous growth as a malignant tumor for over 8 yr was injected into the blastocoel cavity near the inner cell mass of a young mouse embryo. The embryo was then transferred to the uterus of a pseudopregnant (receptive) foster mother of different genetic background. Pregnancy ensued, and a live, normal mouse was born. This mouse was found to be a mosaic, having some tissues derived from the malignant teratocarcinoma, and some from the embryo. Finally, at maturity the mosaic mouse was test mated. The genetic composition of his offspring showed that the genetic material of his sperm was derived at least in part from the embryonal carcinoma cells. A scheme depicting this experimental design is illustrated in Figure 3.5.

The incidence and types of teratomas are species specific and determined by the genetic composition of the organism. In humans for example, ovarian cancer is more frequent than testicular cancer. In the mouse several genetic alleles influence the frequency of this type of cancer, and a particular allele may influence only testicular or ovarian teratocarcinogenesis. In some inbred strains of mice the frequency of teratoma occurrence can be as high as 50%!

The experiments described above reveal the reversible nature of the cancerous transformation process if the cell of origin is an immortal, totipotent germ cell. They illustrate the distinction between growth regulation and immortality. How-

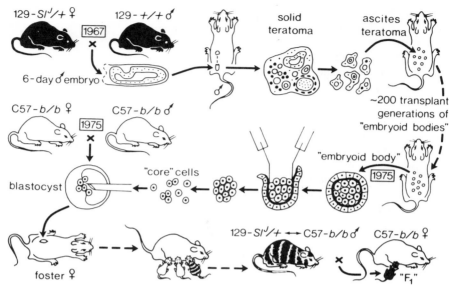

Figure 3.5. Schematic depiction of the experimental design which was used to establish that malignant embryonal carcinoma cells may be genetically normal. These experiments suggested that a nonmutational event is responsible for malignancy in teratocarcinogenesis. The 8-yr history of the experiment is diagrammed, starting at the upper left. The teratoma was experimentally produced in 1967 from a 6-day chromosomally male (X/Y) embryo. The embryo was placed under a testis capsule, where it became disorganized, forming a teratocarcinoma that metastasized to a renal node. The primary tumor was minced and transplanted intraperitoneally; it became an ascites tumor of *embryoid bodies* with yolk sac *rinds* and teratocarcinoma (or embryonal carcinoma) *cores*. In 1975, after almost 200 transplant generations in syngeneic hosts, the rinds of some embryoid bodies were peeled away, and the malignant core cells were injected into blastocysts from parents of a different genetic makeup. The blastocysts were transferred to the uterus of a pseudopregnant foster mother (mated to a sterile vasectomized male). Pregnancy ensued and live normal mice were born. Some had coat-color mosaicism and/or internal tissue contributions of the tumor strain. At maturity, a mosaic male was test-mated to appropriate females. Production of "F_1"-like offspring proved that he produced normal sperm derived from the teratocarcinoma cells. From B. Mintz and K. Illmensee, Proc. Natl. Acad. Sci. USA 72:85–89 (1975), copyright The National Academy of Sciences, reprinted with permission.

ever, passage through a full sequence of linear differentiation steps eventually leads to loss of malignancy, cessation of growth, and eventually necrosis (death). Thus, in teratocarcinogenesis the expression of differentiated traits correlates with loss of unregulated growth characteristics. In the next section we shall examine the second type of cancerous transformation, which dissects growth regulation from expression of a differentiated phenotype. This second transformation process involves events that change the genetic composition of the cell.

Transformation of Normal Somatic Cells

The conversion of a normal tissue cell to a cancer cell probably occurs by a multistep process that affects several different cellular physiological processes. Figure 3.6 illustrates one hypothesis regarding transformation. Normal cells are assumed to be mortal, and their growth is subject to stringent regulatory constraints. Specific mutations may render growth of these cells less sensitive to control by various agents including plasma growth factors and hormones, macromolecules, and other cells with which the developing cancer cells come in direct contact. For example, one mutation may diminish the dependency of the cell on a hormone such as a prostaglandin or insulin, both of which are found in the plasma and are required for the growth of many tissue cells. Subsequent events may abolish cell mortality so that the cell is capable of indefinite growth. These events would be expected to give rise to a malignant cancer.

Some of the evidence leading to the suggestion that all tissue cells are mortal stems from work with cells that have been adapted to the tissue culture environment. Tissue cells can be cultured on plastic plates in a solution containing nutrients, salts, buffers, vitamins, growth factors, and hormones. Such cells will die after 20 to 60 generations, depending on the species, age of the organism,

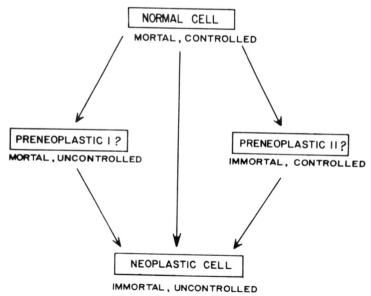

Figure 3.6. Proposed scheme for the stepwise conversion of a normal tissue cell to a malignant cancer cell. Transformation involves both loss of sensitivity to normal growth regulatory constraints and loss of cellular mortality characteristics. This scheme depicts these events occurring sequentially, but it is also possible that a single event (such as virus infection) could effect both changes.

and other factors. When fibroblastic cells of rodents (mice, rats, hamsters) are plated in tissue culture, they proliferate for many generations before growth rates slow and even halt. The vast majority of the cells in the population may die, but a few may survive, remaining in the nonproliferative state for as long as several weeks. Subsequently, cell division again begins, slowly at first, and then increasingly rapidly, until division times of about 1 day are attained. These cells are said to have passed through the *crisis of senescence*. They have been established in culture and are considered to be cell lines that are capable of indefinite growth. Cells that have been exposed to *oncogenic agents* (mutagens that cause cancer) or that are derived from a tumor, however, can frequently be established in culture without passing through the crisis of senescence. These observations lead to the suggestion that mortality is genetically controlled, and that cancerous transformation of a tissue cell can abolish the mortal state. Nomenclature for cancer cells derived from different somatic cell types in the body is illustrated in Figure 3.7.

The preceding discussion suggests that the transformed phenotype of somatic cells is a more or less permanent, heritable characteristic. Recent experiments have led to the conclusion that agents such as certain chemicals, radiation, and *oncogenic viruses* can cause permanent changes in DNA structure (mutations) that may lead to the transformed (immortal) state. Much of this work has focused on understanding the mechanisms by which oncogenic viruses transform their host cells and relating the resulting genetic changes to alterations in the biochemical machinery of the cell.

One such virus, Rous sarcoma virus (RSV), is an RNA-containing virus that has been isolated from chickens. Under appropriate circumstances, this RNA can be copied into a DNA molecule in RSV-infected chicken cells by a viral enzyme called *reverse transcriptase*. This DNA can then become stably integrated into the genome of the cell by recombinational events, where its presence leads to transformation of the infected cell. Such a cell then becomes immortal and is released from contact inhibition of growth (Chapter 10). It passes these characteristics on to daughter cells, and in an intact animal these events can lead to malignant cell growth and death of the organism.

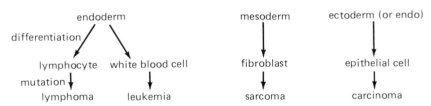

Figure 3.7. Origins and nomenclature of cancer cells derived from somatic tissues within the body of a higher animal. Transformation of somatic cells results from mutational events that render the cell less sensitive to growth regulation. Loss of mortality and differentiation may accompany the transformation process, and multiple mutational events may occur.

The physiological changes that take place in virally transformed cells are numerous and include changes in cellular metabolism, protein and nucleic acid synthesis, and in cell surface antigens. However, in the case of RSV, a mutation in a single viral gene (*src*) renders the virus incapable of transforming host cells. The protein product of this gene has been identified as a 60,000-dalton phosphoprotein that itself is a protein kinase that phosphorylates a number of proteins in the infected cell. These proteins are phosphorylated on tyrosine residues, and ATP is the phosphoryl donor. These observations have led to the hypothesis that transformation by RSV requires phosphorylation of host proteins by the protein kinase encoded by the *src* gene, and that this modification alters the enzymatic and/or structural properties of these proteins. Such essential *viral transforming genes* (*oncogenes*) have now been identified in a number of tumor viruses. Some also encode tyrosine protein kinases as does the *src* gene, while others encode proteins with as yet unknown functions. Table 3.1 lists some of the oncogenes that have been identified so far in RNA tumor viruses that transform their host cells.

Using nucleic acid hybridization techniques and recombinant DNA technology, it has recently been possible to show that normal uninfected animal cells contain DNA sequences that are homologous to many of the viral transforming genes. This has led to the suggestion that tumor viruses may have acquired these genes via recombinational events with normal cellular DNA. Further, it has been suggested that transformation by these viruses may therefore result from overproduction of normal cellular proteins, or from expression of normally repressed cellular genes as a result of attachment to an efficient viral promoter. For example, it has been shown that normal chicken cells have small amounts of the cellular

Table 3.1. Transforming Genes of Acute Transforming Retroviruses

Gene	Prototype Virus	Isolation source
src	Rous sarcoma virus	Chicken
fps	Fujinami sarcoma virus	Chicken
fes	Snyder-Theilin feline sarcoma virus	Cat
yes	Y73 sarcoma virus	Chicken
fms	McDonough feline sarcoma virus	Cat
mos	Moloney sarcoma virus	Mouse
ras	Harvey sarcoma virus	Rat, mouse
sis	Simian sarcoma virus	Woolly monkey
myc	Myelocytomatosis virus strain MC29	Chicken
erb	Avian erythroblastosis virus	Chicken
myb	Avian myeloblastosis virus	Chicken
abl	Abelson leukemia virus	Mouse
rel	Reticuloendotheliosis virus strain T	Turkey
ros	UR2 sarcoma virus	Chicken

Source: From Cooper, G.M. Cellular transforming genes. *Science* 217:801–808, 27 August 1982. Copyright 1982 by the AAAS.

homolog of the *src* gene product, but upon transformation by RSV, the viral gene product is produced in a 100-fold greater concentration.

Most recently, it has been possible to introduce DNA of one animal cell type into another cell type (*transfection*) and to observe its stable integration into the host genome. These experiments have revealed that even DNA from normal cells can, with low frequency, transform certain types of animal cells maintained in tissue culture. Furthermore, molecular clones of normal cell viral oncogene homologs, when attached to viral promoter sequences, have been shown to induce transformation very efficiently in transfection experiments. Finally, it has been demonstrated that within the same differentiated cell type, similar or identical cellular transforming genes are often activated regardless of whether oncogenesis occurred spontaneously or was induced chemically or by tumor viruses.

These observations strongly support the contention that transformation of somatic cells leading to malignancy involves aberrant regulation or overproduction of normal cellular genes in many, if not all, cases. Some of these genes have become part of the genomes of oncogenic viruses, augmenting the reproduction of their nucleic acid through integration and transformation of the host cell. Because many cellular transforming genes have been shown to have been highly conserved throughout evolution, it is likely that they perform essential functions in normal cells. One suggestion consistent with much of the evidence is that these genes, or their protein products, are involved in normal cellular development and differentiation. It is to be expected that the elucidation of the mechanisms by which the expression of these genes is controlled, and of the functions of their products, will profoundly increase our understanding of the molecular events associated with the control of normal progressions of differentiation and with the aberrant behavior that accompanies neoplastic transformation.

Selected References

Bishop, J.M. Enemies within: the genesis of retrovirus oncogenes, *Cell 23*:5 (1981).

Brugge, J.S. and R.L. Erikson. Identification of a transformation-specific antigen induced by an avian sarcoma virus, *Nature (London) 269*:344 (1977).

Collett, M.S., A.F. Purchio, and R.L. Erikson. Avian sarcoma virus-transforming protein, pp60[src] shows protein kinase activity specific for tyrosine, *Nature (London) 285*:167 (1980).

Cooper, G.M. Cellular transforming genes, *Science 217*:801 (1982).

Koprowski, H., ed. *Neoplastic Transformation: Mechanisms and Consequences*, Life Sciences Research Report 7, Dahlem Workshop, Berlin, 1977.

Martin, G.R. Teratocarcinomas as a model system for the study of embryogenesis and neoplasia, *Cell 5*:229 (1975).

Mintz, B. and K. Illmensee. Normal genetically mosaic mice produced from malignant teratocarcinoma cells, *Proc. Natl. Acad. Sci USA 72*:3585 (1975).

Pastan, I. and M. Willingham. Cellular transformation and the "morphologic phenotype" of transformed cells, *Nature 274:*645 (1978).

Saier, M.H. Jr., S. Erlinger and P. Boerner. "Studies on Growth Regulation and the Mechanism of Transformation of the Kidney Epithelial Cell Line, MDCK: Importance of Transport Function to Growth," in *Membranes in Growth and Development,* Alan R. Liss Inc., New York, 1982:

Stiles, C.D., W. Desmond, L.M. Chuman, G. Sato, and M.H. Saier, Jr. Growth control of heterologous tissue culture cells in the congenitally athymic nude mouse, *Cancer Res. 36:*1353 (1976).

Stiles, C.D., W. Desmond, L.H. Chuman, G. Sato, and M.H. Saier, Jr. Relationship of cell growth behavior *in vitro* to tumorigenicity in athymic nude mice, *Cancer Res. 36:*3300 (1976).

CHAPTER 4

Sporulation in Evolutionarily Divergent Prokaryotes and Eukaryotes

> One should not think slightingly of the Paradoxical;
> the paradox is the source of the thinker's passion,
> and the thinker without a paradox is like a lover
> without feeling.
>
> *Kierkegaard*

Evolutionary processes build upon pre-existing genetic material to develop increased degrees of biological diversity and complexity. Thus, the first organisms to appear on earth were undoubtedly very simple, and complexity evolved as dictated by evolutionary pressure for survival. Those organisms that remained in constant environments resembling the primordial seas may have been under little pressure to change, but most organisms were faced with a need to grow, reproduce, and survive in a fluctuating environment which was not always consistent with life as it existed then. In order to obviate the need for constant and favorable external conditions, primitive living organisms followed either or both of two strategies. First, they evolved resting cell structures that could withstand extremes of environmental conditions; and second, they became adhesive, generating the multicellular state so that the intercellular space became an environment which, to some extent, could be controlled by the organism. Among the former groups of organisms are those that produced resting cells of low metabolic activity such as the spores of microbes, seeds of plants, and dormant haploid and diploid eggs of certain animals. Among the latter groups of organisms were animals and plants in which the constituent cells formed a solid mass so that the constitution of the intercellular environment could be regulated by secretory processes. Elements on the external surface of the organism provided protection for cells located internally.

Like plants and animals, microorganisms use both strategies. Numerous species of both prokaryotic and eukaryotic cell structure give rise to resting spores under conditions that are unfavorable to growth. These spores can withstand extremes in environmental conditions. For example, some *Bacillus* spores are resistant to UV irradiation and to temperatures in excess of 100°C. In some organisms (*Bacillus, Saccharomyces*) endospores form within the walls of the mother cell. In other species (*Streptomyces, Neurospora*) the resting cells (conidiospores) result from differentiation of aerial hyphae. Only certain portions of the vegetative cell population sporulate; the remainder form the structural elements of the mycelium. Still other species (*Myxobacteria, Dictyostelium*) form extensive fruiting bodies that house the resting cells. Generally, the more extensive the structure housing the resting cell, the less resistant the spore is to extremes of environmental conditions, presumably because the aerial structures provide protection and prevent exposure to water. *Bacillus* endospores are among the most heat resistant resting cells known; *Streptomyces* conidiospores are considerably more sensitive; and individual myxobacterial spores, present within the protective fruiting body structures, are only slightly more resistant to heat than are the vegetative cells. Aerial hyphae and fruiting bodies form protective structures for housing resting cells just as multicellularity in animals and plants provides homeostatic environments for cell growth. It is worth noting that the myxobacterial fruiting body may also provide a mechanism for initiating cooperative feeding during early vegetative growth following spore germination.

With these relationships in mind, it is reasonable to propose that the two survival strategies employed by living organisms to combat unfavorable environmental conditions co-evolved employing common molecular mechanisms. This argument strengthens the postulate that mechanisms important to microbial sporulation will be found to be operative in one version or another during embryogenesis in humans. It provides incentive for extensive studies into cyclic differentiation in microorganisms. Some of the better-characterized microbial sporulation processes are described in this chapter.

Sporulation in *Bacillus*

Bacillus endospores are probably the most resistant dormant structures of all life forms. They exhibit virtually no oxygen utilization or biosynthetic activity and are viable almost indefinitely. Viable *Bacillus* spores have been recovered from an Egyptian mummy's tomb after a dormancy period in excess of 3000 years!

The structure of a *B. subtilis* spore differs from that of the vegetative cell in several respects (Figure 4.1). While the latter consists of a nucleoid of two genomic copies within a cytoplasm surrounded by a single membrane and wall, the former contains a single copy of the genome present in a relatively dehydrated or gel state surrounded by several protective layers (Figure 4.2). Nearest the nucleoid in the spore protoplast is the cortical membrane, or inner forespore membrane, surrounding an extensive cortical layer that comprises about half of the total dry weight of the spore.

In *Bacillus* spores, the cortex contains a peptidoglycan that differs from that of the vegetative cell in both its amino acid and glycan composition and structure. An unusual spore-specific compound, calcium dipicolinate (Figure 4.3) is found in the spore core in high concentrations (up to 15% of the dry weight of the spore). Other spore-specific, low molecular weight compounds include sulfolactate and 2,3-diphosphoglycerate, both of which are probably also complexed

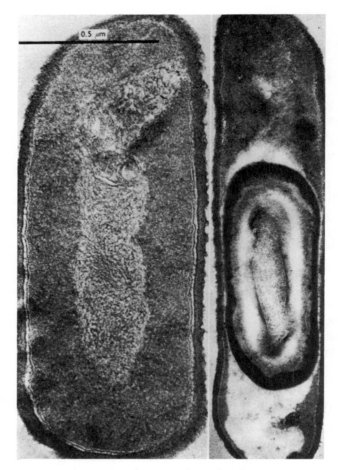

Figure 4.1. Electron micrographs of a vegetative cell (left) and a mature spore (right) of *Bacillus subtilis* illustrating the numerous differences in cell structure. From J. Szulmajster in *Microbial Differentiation,* Symposium 23 of the Society for General Microbiology, J.M. Ashworth and J.E. Smith, (eds.), Cambridge University Press, London, 1973, reprinted with permission.

with divalent cations. External to the cortex is a second membrane, the outer forespore membrane, and two or three spore coat layers. The structural spore coat proteins, which are virtually lacking in the vegetative cell and are encoded by sporulation-specific genes, are rich in cystine and aromatic amino acids and are highly insoluble, hydrophobic molecules. Finally, the outermost layer, the exosporium, contains several unique proteins, and a heteropolysaccharide consisting of a repeating sequence of four sugar residues. Teichoic acids, present in the vegetative cell wall, are absent.

Many enzymes are found in the spore protoplast. Although these enzymes are

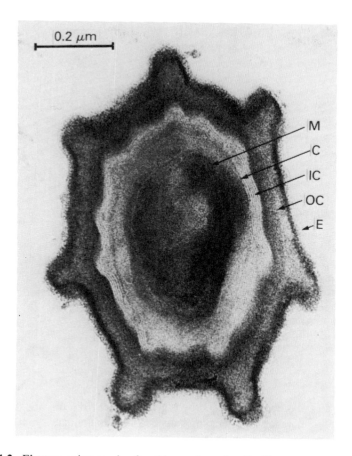

Figure 4.2. Electron micrograph of a thin section of a *Bacillus circulans* endospore showing sculptured ridges in the outer spore coat (OC). IC = layered inner coat; C = cortex; M = membrane; E = exosporium. Reproduced from *Micro-organisms: Function, Form and Environment*, L.E. Hawker and A.H. Linton (eds.), University Park Press (1979), with permission.

newly synthesized during spore differentiation when protein turnover rates are high, most appear to be encoded by the same genes which are expressed in vegetative cells. The spore enzymes, however, are more heat resistant, presumably because they are complexed with protective compounds in a relatively dehydrated state. Some enzymes, such as the enzymes of the tricarboxylic acid cycle, are present in sporulating cells in increased amounts, and mutant bacteria lacking any one of these enzymes cannot sporulate. NADH oxidase, which is membrane-bound in the vegetative cell, is localized to the soluble fraction in

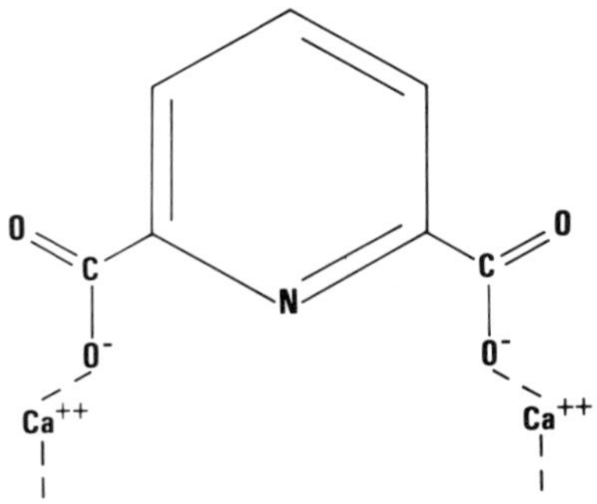

Figure 4.3. Structure of a spore-specific compound, calcium dipicolinate, present in large quantities within the cortex and core of *Bacillus* endospores.

Table 4.1. Unique Biochemical Features of *Bacillus Endospores*

Calcium dipicolinate
Sulfolactate
Spore coat proteins
Unique cortical peptidoglycan
Spore (cortex) lytic enzymes
Antibiotics
Extracellular proteases
A unique NADH oxidase
Altered tRNA patterns
Altered enzyme and metabolite concentrations

the spore and is probably encoded by a gene distinct from the one that encodes the vegetative enzyme. The enzymes needed for dipicolinate synthesis are examples of spore-specific constituents. Unique features of the sporulation process include the production of extracellular antibiotics that kill vegetative cells and proteases that catalyze protein turnover. Table 4.1 lists features which distinguish spores from vegetative cells.

Figure 4.4 illustrates the morphogenic stages of sporulation in *Bacillus*. These morphogenic events are accompanied by biochemical changes that reflect the expression of a temporal program of gene expression. During stage I of the sporulation process (Figure 4.4), after depletion of essential nutrients, synthesis of a number of enzymes is induced. These enzymes include essential catalysts of energy metabolism such as the soluble, spore-specific NADH oxidase and the tricarboxylic acid cycle enzymes. Proteases, ribonucleases, esterases, and phosphatases, also made during the early sporulation stages, probably function in intracellular turnover and are subsequently secreted. During stage II, dipicolinate is made in large quantities, and the developing spores become heat-resistant. It is during stage III, when the double membrane is formed by engulfment of the developing spore by the mother cell membrane, that sporulation becomes irre-

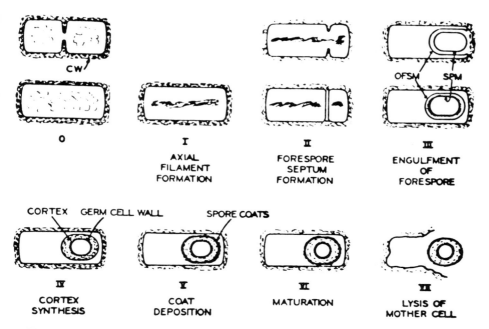

Figure 4.4. Morphogenic stages in the sporulation program of most *Bacillus* species.

versible, and the cell is therefore said to be "committed" to the sporulation program. Coat protein assembly occurs primarily during stage V. Finally, maturation, as a result of the synthesis and assembly of final spore constituents, is accomplished during stages VI and VII.

Over the years, numerous sporulation-specific (*spo*) mutants have been isolated and characterized. Many of these mutants are easily identified visually because they lack the ability to produce melanin pigments and therefore give rise to white colonies. It has been estimated that there must be hundreds of *spo* genes. The different *spo* mutations that have been characterized affect the sporulation process at different morphogenic stages as illustrated in Figure 4.4. While few of the gene products encoded by these genes have been identified, it is clear that some of them function as regulatory components of the differentiation process while others either represent structural constituents of the spore or synthesize spore-specific products. There is little question that a highly regulated and specific temporal sequence of events is responsible for the expression of *spo* genes. While the programmed sequence is initially triggered by adverse environmental conditions, how this temporal sequence is established and regulated is poorly understood. Some recent experimentation bears on this question, however, and one widely accepted hypothesis concerning the mechanism will be presented below.

The principal problem deals with how nutrient deprivation, i.e., limitation for an essential source of either carbon, nitrogen, or phosphorus, can trigger a new program of gene activation that is expressed as a succession of biochemical activities. One can envisage the following sequence: Nutrient deprivation or adverse environmental conditions may alter the concentration of a small regulatory metabolite in the cytoplasm of the bacterial cell. This *second messenger* may then induce synthesis of a new regulatory protein (a *sigma* factor), which interacts with and controls the activity of RNA polymerase. In conjunction with this new regulatory protein, RNA polymerase may catalyze the transcription of sporulation-specific operons. These *spo* genes would code for stage I-specific sporulation proteins, one of which might be a second RNA polymerase regulatory protein that allows expression of stage II sporulation genes. One of the stage II *spo* genes could code for still another sigma factor which switches on expression of a new set of *spo* genes (those specific for stage III). A finite number of steps, determined by a specific number of sigma factors that must act in sequence, would thus determine the temporal pattern of gene expression.

If this hypothesis is even partly responsible for the temporal sequence of events occurring during *Bacillus* sporulation, it can be anticipated that (1) altering the cytoplasmic concentration of a particular metabolite or regulatory molecule may induce sporulation under conditions that otherwise prevent the process, and (2) there must be several different sigma factors interacting with RNA polymerase to influence the frequency with which it transcribes vegetative versus sporulation-specific operons. Current evidence favors the notion that the regulatory molecule controlled by the nutrient supply may be a guanine nucleotide or other guanine containing molecule (possibly GMP) and that this molecule may function in a negative sense. Normally, excess glucose, ammonium, and phosphate repress sporulation. However, when the cellular guanine nucleotide pool is depleted, either by addition of an inhibitor of guanine synthetase or by use of a guanine auxotroph in the presence of limiting guanine, sporulation can be readily demonstrated in the presence of all three nutrients, and the sporulation process is accompanied by all of the biochemical changes discussed above.

With regard to the second possibility, five distinct sigma factors for RNA polymerase have been demonstrated in *Bacillus*. σ^{55} has a molecular weight of 55,000 and only transcribes vegetative genes. σ^{37} is a minor protein of 37,000 dalton size which is probably sporulation specific. σ^{29}, of 29,000 molecular weight, may also exhibit specificity for *spo* genes, transcribing those associated with early-to-intermediate stages of the differentiation process. σ^{32} and σ^{28} are two additional sporulation-specific sigma factors with molecular weights of 32,000 and 28,000 respectively.

A number of *spo* and vegetative genes present on the *Bacillus* chromosome have been cloned, and expression of the cloned genes in the presence of RNA polymerase containing each of three sigma factors has been studied. The results revealed that each of the three factors conferred differing specificities upon the enzyme. Sequence analyses revealed that the consensus promoter sequence for each of the holo RNA polymerases, complexed with each of these sigma factors, differs, but may overlap with those of the other holoenzymes. These results argue in favor of the proposal that a switch in sigma factor synthesis may control the temporal progression of sporulation-specific events.

σ^{55} and σ^{37} have been shown to be present during vegetative growth and disappear during the early phase of sporulation. σ^{29}, on the other hand, is lacking in the vegetative cell but appears during sporulation. One class of sporulation specific mutants (*spoOA*) does not switch on the temporal genetic program, and in these mutants, the new σ^{29} factor does not replace the vegetative sigma factor (σ^{55}). These observations lead to the probability that a switch in sigma factor availability to RNA polymerase is a key feature of the mechanism by which the progression of differentiation-specific events proceeds in a regular linear fashion.

Bacterial Endospore Germination

The process by which a *Bacillus* spore regenerates a vegetative cell under conditions favorable to growth consists of three stages: activation, germination, and outgrowth. Fresh spores do not germinate easily, but any one of a number of treatments "shock" or activate them so that germination is facilitated. Aging for several weeks, sublethal heating, or exposure to low pH, sulfhydryl-containing compounds, or organic solvents such as dimethylformamide serve to activate the spore. Activation is required for fresh bacterial endospores, but not for germination of bacterial or eukaryotic conidiospores or spores derived from fruiting bodies.

Numerous agents have been found to initiate germination of activated endospores, and these are to some extent species-specific. It is possible to classify the initiators in *Bacillus* as follows: (1) Nutrients such as sugars, amino acids, and nucleotides can serve as initiators. In some *Bacillus* species a nonmetabolizable sugar analogue can initiate germination, indicating that initiation by a sugar does not necessarily involve energy generation. In most species of *Bacillus*, L-alanine is a potent germinant, while adenosine potentiates germination. (2) A non-nutrient chemical germinant specific to bacterial endospores is calcium dipicolinate, and spores deficient for this compound germinate poorly. (3) Lysozyme and spore lytic enzymes can initiate germination, particularly if the spores have been previously treated with reagents that alter or remove the coat. (4) Finally, there are physical initiators that probably function (as do the lytic enzymes) to disrupt the outer coat layers, thereby facilitating release of the protoplast as a new vegetative cell.

Germination of a *Bacillus* spore consists of an ordered sequence of measurable events (Table 4.2). During this process a major part of the dry weight of the spore (up to 50%) is lost. These materials include the cortex and cortical substances, proteins, peptides, and calcium dipicolinate, and about 80% of the exosporium. The remainder of this germ cell layer appears to be incorporated into the developing vegetative cell.

Finally, in the presence of sufficient sources of nutrients, outgrowth follows germination. The morphological events accompanying outgrowth include swelling of the protoplast to 2–3 times its original volume, elongation to a rod-shaped

Table 4.2. Sequence of Events Occurring During Germination of *Bacillus* Spores

1. Loss of heat resistance
2. Acquisition of chemical sensitivity
3. Release of K^+ and Ca^{++}
4. Release of dipicolinate
5. Phase contrast darkening
6. Degradation and loss of cortical materials
7. Dramatic fall in the absorbance of a spore suspension.

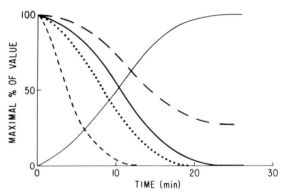

Figure 4.5. Time course for the biochemical changes which accompany outgrowth of a *Bacillus* spore. --- Heat resistance. ··· Internal $[Ca^{24}]$. ——[3-phosphoglycerate]. —— Absorbance. —— [ATP].

cell, and finally, septation. Approximately 200 min are required to achieve the first cell division event after addition of a germinant. Figure 4.5 illustrates the sequential appearance of certain biochemical events that accompany the outgrowth process. These and other events include (in order of occurrence): RNA and protein synthesis, appearance of certain enzyme activities and restoration of respiration, cell wall synthesis, DNA synthesis, and cell division.

Ascospore Formation in Yeast

Examination of sporulation in yeast and bacteria reveals many similarities in the morphological and biochemical changes that occur in response to adverse physiological conditions. Yeast spores, like those of *Bacillus* and *Clostridia,* originate as endospores within the parental cell. The major differences between yeast and bacterial sporulation processes result from the two facts that the former organism is a eukaryote with functionally specialized organelles lacking in bacteria, and that yeast sporulation is sexual and normally involves meiotic reduction from the diploid to the haploid state while bacteria are already haploid. Consequently, most yeast species, including *Saccharomyces,* generate an *ascus* containing four *ascospores*. However, some yeast species generate only one spore per parental cell as do Bacilli, while others usually produce 2 or 8 spores per ascus. The protoplast of the yeast ascospore is about 10- to 100-fold larger than the bacterial spore protoplast, and it houses all essential eukaryotic organelles including a nucleus, endoplasmic reticulum, and mitochondrial granules.

As in bacteria, the yeast spore wall consists of outer and inner spore coats. The outer coat is rich in protein while the inner coat, like the vegetative cell wall, consists of a rigid glucan layer complexed with protein and mannan. The hydrophobicity of this layer is due to its high lipid content (as for Clostridial spores) rather than to the hydrophobicity of coat proteins (as for *Bacillus* spores). Like bacterial spores, yeast ascospores can be maintained in a dormant state for decades, and they are more resistant to heat, alcohol, and cell wall lytic enzymes than are vegetative cells. They are, however, much more sensitive to these agents and treatments than are *Bacillus* spores.

Physiological conditions that promote sporulation in *Bacillus* are similarly effective in *Saccharomyces*. Thus, limitation of essential nutrients, an adequate source of oxygen, and an oxidizable substrate such as acetate are prerequisites for sporulation. Glucose, various amino acids, and other rich sources of nutrition repress sporulation. In this regard, it is interesting that in *Saccharomyces,* as in *Bacillus,* guanine nucleotides may play a role as second messengers in signaling conditions appropriate for sporulation.

During yeast sporulation a prospore wall appears around each lobe of the four-lobed nucleus within the cell. The nuclear membrane, which will comprise two spore-delineating membranes, spreads around each nuclear lobe and finally joins when the lobe separates to form the spore nucleus. Subsequently, cell wall materials are deposited between the two membranes in a process reminiscent of the formation of the spore protoplast and cortical layer in *Bacillus*. In *Saccharomyces* as in *Bacillus,* sporulation is accompanied by an increase in the rates of RNA and protein turnover, and proteases and nucleases involved in their breakdown are eventually released from the ascus when it ruptures, liberating the ascospores. Transcription and translation occur continuously throughout sporulation both in the nucleus and in the mitochondria. Among the proteins which are derepressed during sporulation are the tricarboxylic acid cycle enzymes and those of the glyoxylate shunt. A host of new spore-specific products are also synthesized. While some of these differ from those made during sporulation in *Bacillus,* their functions undoubtedly overlap. Thus, in spite of some differences in structure, composition, and physiological properties of bacterial and yeast spores, the physiological and biochemical changes accompanying spore formation exhibit numerous similarities. It seems reasonable to suggest that a single fundamental mechanism of spore development and maturation is operative, but that this mechanism is modified in order to accommodate the meiotic process in yeast as well as differences in cell structures.

Conidiation in *Streptomyces* and *Neurospora*

Among the sporulating bacteria are found a rather unique group of organisms, the *Actinomycetes*. These bacteria produce multinucleoid, branching mycelia, some parts of which play vegetative roles while other parts play reproductive roles. Superficially these prokaryotes resemble eukaryotic fungi; in fact, they were originally classified as fungi before detailed examination revealed bacterial ultrastructure and peptidoglycan-containing cell walls. Genetic exchange and phage susceptibility are also typically bacterial. A close examination of the processes of conidiation in *Streptomyces* and *Neurospora* reveals striking similarities suggesting that they may have a common evolutionary origin.

Different genera within the *Actinomycetes* produce different kinds of spores. *Thermoactinomyces* produce aerial hyphae bearing heat-resistant polyhedral *endospores* arranged singly on hyphal protrusions (Figure 4.6A). *Dermatophilus*, on the other hand, generates hyphae that subdivide longitudinally and transversely into compartments housing spores which, when released, are motile (Figure 4.6B). In *Streptomyces,* chains of conidiospores are produced following subdivision of the long apical cells of the aerial branches (Figure 4.6C). The conidiospores of various *Streptomyces* species are quite striking in appearance as revealed in Figure 4.7. The morphologies of conidia of the various fungal species are equally varied.

Spore germination, vegetative growth, production of aerial hyphae and conidiation in *Streptomyces* species all resemble these same processes in *Neurospora crassa* to a remarkable degree. Upon encountering an environment favorable to growth, a *Streptomyces* or fungal spore will germinate. It swells and loses refractility, reducing power is generated, and cell wall lytic enzymes are activated. Action of the lytic enzymes allows germ tube emergence, and the growing germ tube develops into vegetative, substrate hyphae. In *Neurospora* certain compounds such as cysteine and ethylene glycol prevent germ tube generation without inhibiting germination. The volume of the spherical cell may increase ten fold,

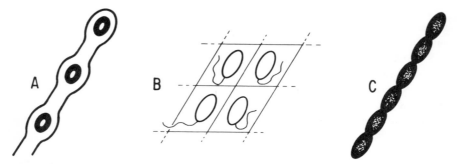

Figure 4.6. Depiction of different spore types made by different genera of *Actinomycetes*. (A) Endospores of *Thermoactinomyces* arranged singly on hyphal protrusions; (B) Motile spores of *Dermatophilus* housed in hyphal compartments; (C) Chains of conidiospores of *Streptomyces*.

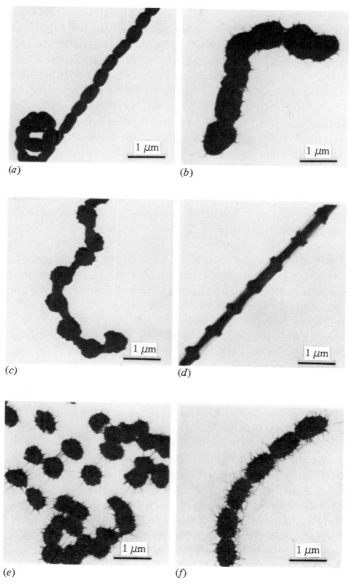

Figure 4.7. Morphologies of conidiospore chains produced by different species of *Streptomyces*. Courtesy of H.D. Tresner, Lederle Laboratories; reproduced in part from E.B. Shirling and D. Gottlieb, Cooperative Description of Type Cultures of *Streptomyces*. II. Species Descriptions from First Study. *Intern. J. Syst. Bacteriol. 18*:69–189 (1968).

cell mass increases, DNA replication occurs, and net RNA and protein synthesis is initiated normally. Possibly these inhibitors prevent localization of the cell wall lytic enzymes (the "can opener"), which initiate germ tube emergence.

Following germination of a *Streptomyces* or *Neurospora* spore and during the first 24 to 48 hr of clonal growth on solid medium, the major hyphae branch at intervals, thus producing increased numbers of substrate hyphae as the colony increases in size. The hyphal diameter of the bacterium is about 1 μm, while that of the fungus is about 5 μm. Transverse septa develop in the older portions of the mycelia of both organisms, but growth is confined to apical regions. In *S. coelicolor,* side branches emerge from older, non-growing hyphae. In *Neurospora,* branching, like growth, usually occurs at the apical tip. Branching from older mycelia is a relatively rare event in the fungus.

After about 2 days in culture, a colony of *Streptomyces coelicolor,* initially shiny and "bald," begins to develop aerial mycelia. The aerial hyphae originate as vertical branches from the substrate mycelium and give the colony a hairy white appearance. They resemble the substrate mycelia in having a diameter of about 1 μm, but are straighter and less branched. Interestingly, the walls of the aerial mycelia are of a different composition from the substrate mycelia and exhibit a markedly hydrophobic character. This characteristic presumably prevents desiccation and facilitates the passage of nutrients through the aqueous cytoplasm from the substrate mycelia to the growing aerial tip.

Under similar culture conditions, wild type *N. crassa* will begin to develop aerial hyphae after about 24 hours following germination. Like those of the bacterium, the aerial hyphae develop as vertical branches from the substrate mycelia. They are less broad than substrate mycelia having a diameter of 2–3 μm. Moreover, the compositions of aerial and substrate mycelia differ as do those of *S. coelicolor.* Because septum formation is incomplete, cytoplasmic streaming and nutrient flow from the substrate mycelia to the aerial tip occurs unimpeded.

After extension of aerial hyphae, the *Streptomyces coelicolor* colony turns from white to gray as sporulation occurs. Several steps are involved as shown in Figure 4.8. First, the hyphal tips coil and sometimes fragment. Second, the long apical cell (above the last crosswall) is subdivided by the almost simultaneous appearance of a number of closely spaced sporulation septa. These differ markedly from the transverse septa found in the remainder of the substrate and aerial hyphae, and their formation involves distinct genes as revealed by genetic analyses. Third, the developing spores become more spherical, and a thick spore wall is laid down. Finally, the spore matures, and as it does so, it becomes pigmented, more ellipsoid, and the old cell wall, which is external to the spore wall, disintegrates. Disintegration of the cell wall leaves intact the characteristic chains of mature conidiospores as shown in Figure 4.7, ready for dispersal.

Similar steps are involved in the *Neurospora crassa* conidiation process with some exceptions as shown in Figure 4.9. First, the hyphal tips begin to form lateral branches of anucleate conidiophores, and nuclei migrate asynchronously to the conidiophores. Second, the nucleated conidiophores swell from a diameter

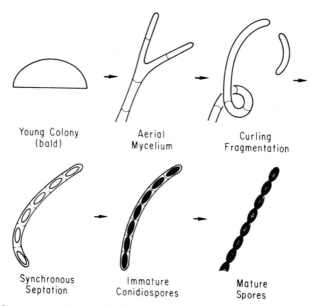

Figure 4.8. Stages in the conidiation processes occurring in *Streptomyces coelicolor*.

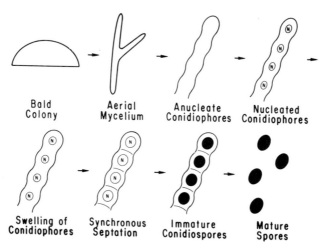

Figure 4.9. Stages in the conidiation process occurring in *Neurospora crassa*.

of about 3 μm to a diameter of 4–6 μm and then bud to become spheres as shown in Figure 4.10. Third, complete septation occurs by a synchronous mechanism. These sporulation septa differ from those in the substrate and aerial hyphae, and their formation involves different genes. Mutations affecting conidial septa do not alter vegetative septa and vice versa. Finally, the spore matures while the external cell wall disintegrates by a lytic process. The mature conidiospores are ready for dissemination.

A number of developmental mutations in *S. coelicolor* have been characterized that allow normal rates of vegetative mycelial growth, but interfere either with aerial hyphae production or with conidiation. Those that cannot form aerial mycelia are designated *bld* (for bald) while those that cannot sporulate are designated *whi* (for white). Among the latter mutations are some that arrest development at each of the stages depicted in Figure 4.8. For example, different classes of *whi* mutants are defective for coiling, for sporulation septation, for immature spore formation, or for the terminal step of maturation. The genetic analyses clearly show that distinct genes code for sporulation-specific functions in each of the developmental stages shown in Figure 4.8, and that these genes comprise part of a program which, though essential for sporulation, is not required for normal vegetative growth. Genetic analyses of analogous mutations in *Neurospora crassa* lead to exactly the same conclusions, thus emphasizing the probable similarities between the genetic regulatory mechanisms controlling hyphal growth and conidiation in these two evolutionarily divergent organisms.

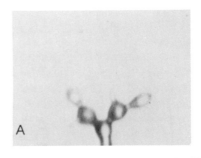

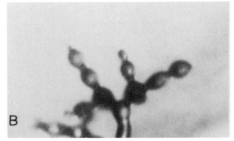
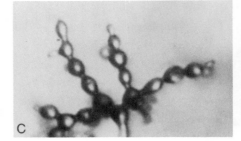

Figure 4.10. Time-lapse sequence of conidiophore formation by repeated budding in *Neurospora crassa*. From S.S. Matsuyama, R.E. Nelson and R.W. Siegel, *Devel. Biol. 41*:278–287 (1974), copyright Academic Press, reprinted with permission.

Fruiting in *Myxobacteria* and *Dictyostelium*

Myxobacterial species, sometimes referred to as slime bacteria, produce fruiting bodies of differing degrees of complexity. In some (*Cystobacter*), fruiting body sporangia form on the agar surface without an appreciable stalk. *Myxococcus* generates a single nonbranching sporangium on a short stalk (Figure 4.11A); *Stigmatella* species form sculptured fruiting bodies with discrete sporangia on a thick branching stalk (Figure 4.11B); while *Chondromyces* produces branched fruiting bodies with apical cysts that house the myxospores (Figure 4.11C). A photograph of a *Chondromyces* fruiting body is shown in Figure 4.12A. This is the most complex of all fruiting bodies constructed by the myxobacteria. These structures reach maximal heights of about 1 mm and are about the same size as those of *Dictyostelium*. Germination in *Chondromyces* results from exposure of the moist cysts and myxospores to nutrients. Usually the spores germinate within the cyst, the cyst ruptures, and a swarm of vegetative bacteria is released (Figure 4.12B). These cells then employ *gliding* motility to migrate as a cohesive pack in search of food sources. When two such swarms of gliding cells come close to one another, they merge, probably as a result of strong chemotactic attractive forces.

The most widely studied myxobacterial species is *Myxococcus xanthus*. These bacteria divide with a maximal generation time of about 4 hr. They can be grown both in liquid and on solid media, and a growth medium of defined composition is available. *M. xanthus* cells can be easily cloned. A variety of developmental mutants are available, and genetic analyses are possible because transduction in myxobacteria has been demonstrated with the myxococcal phage MX3 as well as the *E. coli* phage, P1. Genetic analyses can also be conducted using transposons. Synchronous sporulation can be studied in liquid culture in the absence of fruiting by addition of high concentrations of glycerol. However, the spores generated in solution never attain the same degree of dormancy as those formed within a fruiting body, and they differ from the latter in having thin coats and lacking several spore-specific proteins.

Like the myxobacteria, the different species of slime molds form fruiting bodies of varying morphologies. Most are simple, with a single sporangium on

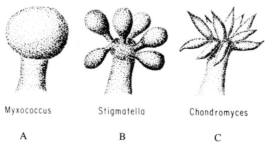

Myxococcus   Stigmatella   Chondromyces

A   B   C

Figure 4.11. Schematic depiction of fruiting bodies formed by different myxobacterial genera: (A) *Myxococcus;* (B) *Stigmatella;* and (C) *Chondromyces.*

A

Figure 4.12. Photograph of a fruiting body of *Chondromyces appiculata* (A) and of germinated myxospores, released from a moistened cyst (B). From J.W. Wireman and M. Dworkin, *Science 189:*516–523 (1975), copyright 1975 by the AAAS, reprinted with permission.

an extended stalk, but a few species give rise to branching sporangia. Under favorable conditions the spores germinate with the emergence of solitary amoebae that migrate by means of amoeboid motion over the surface of the substratum in search of food sources. As for the myxobacteria, their natural food sources are live microorganisms.

The most detailed studies of development within the slime molds have been performed with *Dictyostelium discoideum*. The vegetative amoebae divide by

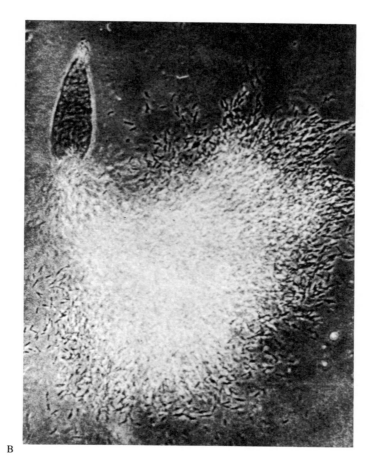

B

Figure 4.12 (continued).

mitosis with a generation time of about 24 hr and can be grown in liquid as well as on solid nutrient media. Mutant strains are easily isolated, and many developmental mutants have been analyzed physiologically, genetically, and biochemically. Consequently, the technical approaches available for studies of bacterial development can be employed in studying slime mold development.

The developmental life cycles of the cellular slime molds and the myxobacteria are superficially similar (Figures 4.13A and B, respectively). In the presence of a sufficient supply of an appropriate bacterial, algal, or fungal food source, both organisms grow vegetatively. Upon exhaustion of the food source, aggregation is initiated. In *Dictyostelium,* a sheath-enclosed aggregate of cells, a *slug,* then migrates in response to light and warmth. Similarly, aggregated myxobacteria can swarm for considerable periods of time after exhaustion of the food source until a suitable site for fruiting is found. The migrating slug or swarm of cells

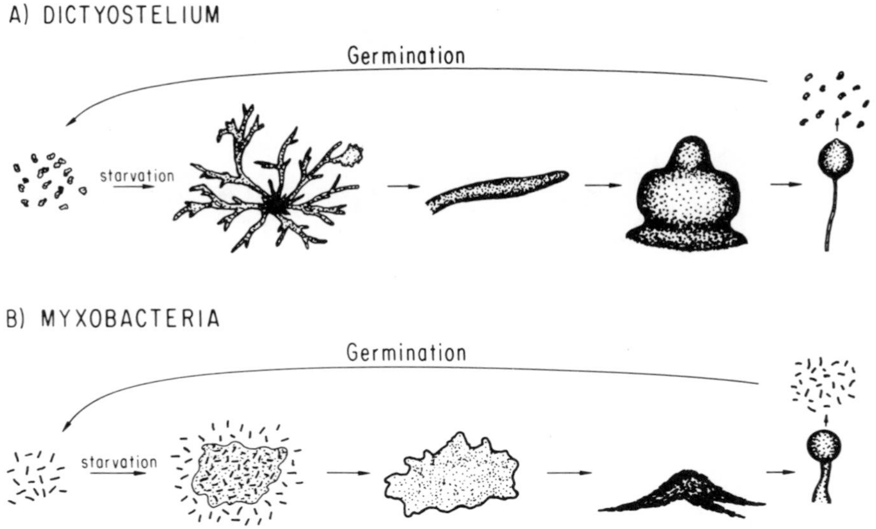

A) DICTYOSTELIUM

B) MYXOBACTERIA

Figure 4.13. Steps in the differentiation cycles of the cellular slime molds (A) and the myxobacteria (B).

comes to rest and culminates with the construction of a fruiting body in which the stalk, consisting of a hardened macromolecular matrix and dead cells, supports spore-filled sporangia. In both cases, those cells destined to form the stalk are "programmed" to die. The similarities of the prokaryotic and eukaryotic fruiting processes are evident from a cursory examination of Figure 4.13.

Closer examination of each of the developmental stages reveals further similarities, but also some differences between the physiological behavior of these two groups of organisms. During vegetative growth, when they are growing on solid media, both groups of organisms grow by lysing and digesting food bacteria or eukaryotic microorganisms by mechanisms involving secretion of lytic enzymes. In the case of the myxobacteria, the digestive enzymes are secreted into the extracellular medium so that lysis of the food microbes can occur before the predator comes in contact with the prey. It may be because of the essential role of extracellular hydrolases that myxobacteria remain social throughout vegetative growth. By contrast, the cellular slime molds always feed as solitary amoebae. In *Dictyostelium,* the digestive enzymes are not secreted into the extracellular medium but instead are transported to lysosomal vesicles. Live microorganisms are swallowed by phagocytosis, and cell lysis and partial digestion are accomplished within the phagocytic vesicles (secondary lysosomes). The occurrence of digestion within an intracellular compartment abolishes the need for high concentrations of extracellular hydrolases (apparently required for digestion in some myxobacterial species), and allows these vegetative cells to exist in the solitary (asocial) state.

Myxobacteria and the cellular slime molds are referred to as slime bacteria and slime molds, respectively, because these organisms secrete an extracellular slime layer. In both groups of organisms the constituents of slime are synthesized during cellular migration following deprivation of a food source. However, slime secretion is dramatically induced upon starvation of *Dictyostelium,* particularly during slug migration, while copious amounts of slime are secreted during vegetative growth of many myxobacterial species. It has been suggested that the myxobacterial slime provides not only traction and a liquid substrate through which the bacteria glide but also a medium through which chemotactic signals can be transmitted. However, a chemotactic mode of directed movement has not been conclusively demonstrated in myxobacteria, and it is possible that tactic behavior is mediated by direct cell–cell or cell–substratum contacts. The slime layer of the *Dictyostelium* slug clearly functions to propel the organism forward as this layer is stationary on the substratum, and the cells move forward only as fast as new sheath is synthesized. Slime does not serve as a medium for transmission of chemotactic signals. It is interesting to note that the motility of both classes of organisms is temporarily blocked when the cells are dividing.

In both groups of organisms, the trigger for aggregation is the same: starvation. In *Dictyostelium,* exhaustion of nutrient supplies induces enzymatic machinery for the emission and reception of chemotactic signals. While this chemotactic process is well characterized and the principal attractant is known to be cyclic AMP, there is considerable uncertainty about the myxobacterial process, in which no attractant has been identified. It is interesting to note that both classes of organisms show oscillatory wave behavior when examined on a macroscopic scale. This behavior suggests elaborate relay systems, possibly involving a similar mechanism.

Extracellular signaling molecules, pheromones or morphogens that influence the development of cellular cohesiveness and fruiting body formation, have been identified for both myxobacteria and the slime molds. As indicated in Figure 4.13, the aggregated prokaryotic and eukaryotic swarms may migrate as a *community* or *organism* for hours before coming to rest. While the cellular slime mold slugs exhibit clear photo- and thermotaxis, allowing them to seek an appropriate environment for culmination, it is not known whether myxobacterial swarms respond similarly. Further studies will be required to determine whether or not they possess this capability.

The eukaryotic fruiting process has been studied in far greater detail than the corresponding process in bacteria. The eukaryotic extracellular stalk matrix is known to consist largely of polymerized cellulose, and stalk cells vacuolize and differentiate before they die (see Chapter 5). The extracellular matrix of the myxobacterial stalk has not been characterized, and the dead cells found within this structure do not differ markedly in morphology from the vegetative cells. Because the two groups of organisms produce fruiting bodies that are similar in size, shape, and pigmentation (red, orange, yellow, or green), it is sometimes difficult, without careful examination, to know whether a fruiting body encountered in nature is of prokaryotic or eukaryotic origin.

An interesting question regarding the evolution of the fruiting processes in these divergent classes of organisms concerns whether *divergent* or *convergent* evolution has occurred, i.e., whether or not common ancestral genes were utilized. Investigators favoring the concept of convergent evolution suggest that similar habitats and life styles led to parallel, separate, and convergent patterns of cellular organization. Those arguing for divergent evolution suggest that genes were exchanged or derived from a common ancestral source early in the evolutionary scheme. The similarities among these two classes of organisms seem to suggest that at least some components of the developmental programs were shared during evolution. Further support for this notion has come from protein-sequencing data, which show that numerous enzymes in prokaryotes share extensive sequence homology with enzymes in eukaryotes. For example, a myxobacterial serine protease exhibits extensive sequence homology with certain animal serine proteases, while part of a development-specific protein in *Myxococcus xanthus*, (called the S-protein) has the same sequence as the active site of a eukaryotic calcium-binding protein. Prokaryotic and eukaryotic ion translocation ATPases have also been shown to exhibit sequence homology. These observations clearly point to a common pool of genetic material in divergent groups of organisms, but do not rule out convergent evolution. Common environmental pressures may have provided the driving force for expression of shared silent genetic material encoding analogous programs of differentiation. Organisms that share a common habitat are also more likely to exchange genetic material than those that do not. Convergent and divergent evolutionary schemes should not be considered mutually exclusive.

Selected References

American Society of Zoologists, *Principles and Problems of Pattern Formation in Animals,* Symposium of the Division of Developmental Biology, 1980, *American Zoologist,* Vol. 22, 1982.

Ashworth, J.M. and J.E. Smith, eds. *Microbial Differentiation,* Society for General Microbiology Symposium 23, Cambridge University Press, 1973.

Brody, S. "Genetic and Biochemical Studies on *Neurospora* Conidia Germination and Formation" in *The Fungal Spore: Morphogenetic Controls* (G. Turian and H.R. Hohl, eds.), Academic Press, London, 1981.

Dworkin, M. "Spores, Cysts, and Stalks" in *The Bacteria,* Volume VII, Mechanisms of Adaptation (I.C. Gunsalus, J.R. Sokatch, and L.N. Ornston, eds.), Academic Press, New York, 1979.

Dworkin, M. Tactic behavior of *Myxococcus xanthus, J. Bact., 154:*452–459 (1983).

Fowell, R.R. "Ascospores of Yeasts" in *Spores VI* (P. Gerhardt, R.N. Costilow, and H.L. Sadoff, eds.), American Society for Microbiology, 1975.

Gerhardt, P., R.N. Costilow, and H.L. Sadoff, eds. *Spores VI,* 6th International Spore Conference, American Society for Microbiology, 1975.

Gould, G.W. and A. Hurst, eds. *The Bacterial Spore,* Academic Press, London, 1969.

Johnson, W.C., C.P. Moran, Jr., and R. Losick. Two RNA polymerase sigma factors from *Bacillus subtilis* discriminate between overlapping promoters for a developmentally regulated gene, *Nature 302:*800–804 (1983).

Kaiser, D. and C. Manoil. Myxobacteria: cell interactions, genetics, and development, *Ann. Rev. Micro. 33:*595 (1979).

Leighton, T. and W.F. Loomis, eds. *The Molecular Genetics of Development,* Academic Press, New York, 1980.

Levinson, H.S., ed. *Sporulation and Germination,* 8th International Spore Conference, American Society for Microbiology, 1981.

Loomis, W.F. *Dictyostelium discoideum, A Developmental System,* Academic Press, New York, 1975.

Loomis, W.F., ed. *Development of Dictyostelium,* Academic Press, New York, 1982.

Losick, R. and J. Pero. Cascades of sigma factors, *Cell 25:*582 (1981).

Matsuyama, S.S., R.E. Nelson, and R.W. Siegel. Mutations specifically blocking differentiation of macroconidia in *Neurospora crassa, Devel. Biol. 41:*278 (1974).

Parish, J.H., ed. *Developmental Biology of Prokaryotes,* Studies in Microbiology, Volume 1, University of California Press, Berkeley, 1979.

Schmit, J.C. and S. Brody. Biochemical genetics of *Neurospora crassa* conidial germination, *Bact. Rev. 40:*1 (1976).

Stephens, K., G.D. Hegeman, and D. White. Pheromone produced by the myxobacterium *Stigmatella aurantiara, J. Bact. 149:*739–747 (1982).

Whittenbury, R. and C.S. Dow, Morphogenesis and differentiation in *Rhodomicrobium vannielii* and other budding and prosthecate bacteria, *Bact. Rev. 41:*754 (1977).

Young, M. Bacterial endospore development—an ordered sequence of gene transcription, *TIBS 3:*55 (1978).

CHAPTER 5

Control of Spatial Differentiation by Chemical Morphogens

> Archetypal ideas have their origin in the archetype, which in itself is an irrepresentable, unconscious, pre-existent form that seems to be part of the inherited structure of the psyche and can therefore manifest itself spontaneously anywhere, at any time. Because of its instinctual nature, the archetype underlies the feeling-toned complexes and shares their autonomy.
>
> C. G. Jung

In the developing human embryo a single cell, the fertilized egg, gives rise to a multitude of cell types through a program of sequential but branching developmental steps. The differentiating cells must not only attain their ultimate morphologies and biochemical properties, they must also assume their proper positions within the developing embryo.

Numerous transplantation studies have established the importance of position to the ultimate fate of an uncommitted or partially committed stem cell. Even in "randomized" differentiating systems such as a teratocarcinoma, discussed in Chapter 3, embryonic inductive effects are revealed by the clustering of cell types with complementary functions. One must superimpose an understanding of positioning, or spatial differentiation, upon the concept of temporal differentiation in order to comprehend fully the regulatory constraints that must be operative during embryogenesis.

In the previous chapter, several organisms that undergo spatial as well as temporal differentiation were discussed. In these organisms a single cell type gives rise to at least two distinct cell types with different structures; the positioning of these two cell types relative to one another is predetermined. In this chapter, the evidence dealing with the mechanisms of positional morphogenesis will be considered for three microbial systems: *Dictyostelium*, *Anabaena*, and *Hydra*.

As discussed in Chapter 4, exposure of growing amoebae of the cellular slime mold, *Dictyostelium discoideum*, to starvation conditions causes the individual amoebae to swarm to an aggregation center that subsequently secretes a slime sheath, encompassing the aggregated cells in a rudimentary multicellular organism. This slug can migrate as a unit for hours. When it finally comes to rest to erect a fruiting body, the cells within the slug differentiate into two different positionally predetermined cell types.

A second microbial example in which a single cell gives rise to two distinct cell types in a spatially regulated fashion is provided by the blue-green bacteria (Cyanobacteria). In species of *Anabaena,* for example, certain cells within chains of photosynthetic, vegetative cells, when deprived of (starved for) an organic nitrogen source, differentiate into thick-walled nitrogen-fixing heterocysts. The positioning of the heterocysts along the chain of vegetative cells is finely regulated.

Hydra provides still a third biological system in which fairly detailed information concerning positional morphogenesis is available. While in all three systems, chemical *morphogens* are thought to influence the ultimate position of the developing cell type, in *Hydra,* one of the morphogenetic agents has been isolated and characterized. We shall discuss all three of these systems partly to illustrate the application of similar principles to evolutionarily divergent biological systems, but also to allow formulation of a molecular mechanism for spatial differentiation. It will be seen that an understanding of spatial differentiation in microbial systems leads to some specific predictions regarding possible mechanisms of embryonic induction.

Spatial Control of Fruiting Body Formation in *Dictyostelium*

The cellular slime molds can exist either as solitary amoebae or as multicellular structures. Amoebal aggregation gives rise to a sheath-enclosed multicellular slug or plasmodium that can migrate in response to light and warmth to an appropriate site for fruiting. Culmination consists of the construction of a spore encasement (sorogen) mounted on a stalk of inviable cells (Figure 5.1). Thus, cells within the slug may give rise to small, dehydrated, dormant spores that will provide the viable element of the mature fruiting body, or to large, vacuolated, cellulose-producing stalk cells that are destined or "programmed" to die.

The fate of a particular cell is determined during slug migration. If left unperturbed, cells comprising the anterior 20% of the slug (prestalk cells) will give rise to vacuolated stalk cells while those comprising the posterior 80% (prespore cells) will differentiate into dormant spores. These two cell types differ biochemically, synthesizing different proteins. For example, prestalk cells make large quantities of actin, which is essentially lacking in prespore cells, while the latter cells synthesize spore coat proteins and mucopolysaccharide-biosynthetic enzymes that are not found in prestalk cells. The two cell types also differ in their adhesive and chemotactic properties so that randomly mixed cells will sort themselves out with prestalk cells separating from prespore cells. These properties presumably keep the two cell types from mixing or streaming during slug mi-

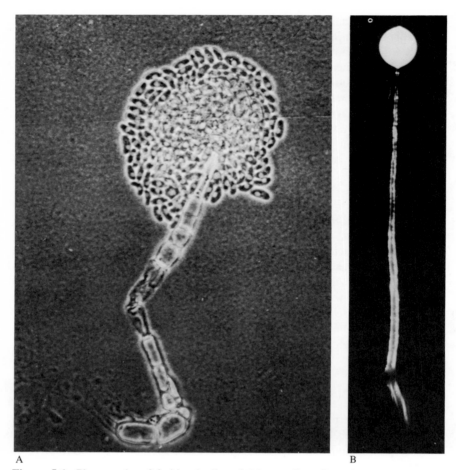

A   B

Figure 5.1. Photographs of fruiting bodies of *Dictyostelium discoideum* consisting of 122 cells (A) or about 100,000 cells (B). The proportion of spore cells (in the sorogen) to stalk cells (providing support for the sorogen), is largely size invariant. Figure A illustrates the major size and structural differences between the two cell types. (A) From W.F. Loomis (ed.), *Development in Dictyostelium* (1982), copyright Academic Press, reprinted with permission; and (B) from J.H. Morrissey, P. Farnsworth, and W.F. Loomis, *Devel. Biol. 83*:1–8 (1981), copyright Academic Press, reprinted with permission.

gration and allow for the permanence of the *fate map* (Figure 5.2). The structural differences between the two cell types in a mature fruiting body are illustrated in Figure 5.1.

The ratio of prestalk to prespore cells within a migrating slug is almost size-invariant, being approximately 1:5 over a 1000 fold size range (Figure 5.1). Although this relationship is highly regular and reproducible, determination is

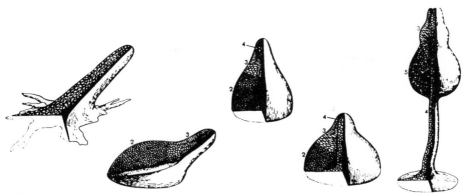

Figure 5.2. Illustration of the "fate map" of cells during slug migration and subsequent culmination. At the end of aggregation all cells appear the same (1), but in the slug they are of two types (2 and 3). The cells near the tip (3) gradually turn into stalk cells (4) and move down inside the mass. The others (2) become spores (5) as the growing stalk lifts them into the air. From Differentiation in social amoebae, J.T. Bonner. Copyright © 1959 by Scientific American, Inc. All rights reserved.

fully reversible during slug migration. However, the rate of the prestalk to prespore cell transition differs from that of the prespore to prestalk cell transition: Prespore isolates induced to culminate immediately after separation of the two cell types form fruiting bodies of normal proportions, but freshly isolated prestalk cells induced to culminate immediately give rise to a spore-free stalk. Only if the isolated anterior portion of the slug is allowed to migrate for a day or two does culmination give rise to a fruiting body of normal proportions.

The cells within the migrating slug continually secrete slime material, which forms a sheath encasement for the amoebae. The slime sheath is stationary on the substratum and may provide traction for the motile cells within. As the cells migrate out of the old sheath, it is left behind as a collapsed tube. Because the sheath is synthesized along the entire length of the slug, it is thinnest at the anterior tip, becoming progressively thicker as it passes over the more posterior regions. It is therefore reasonable to assume that the sheath is more permeable to diffusible molecules at the anterior tip and less permeable in the posterior regions of the slug. It has been suggested that this property may provide a molecular explanation for prestalk and prespore determination during slug migration.

If we assume that morphogenetic substances quantitatively influence development of an amoeba into a prestalk or prespore cell, and that one (or more) of the morphogenetic substances is a small diffusible molecule, then the concentration of the substance within different portions of the slug will be determined in part by the diffusibility of the substance across the slime sheath. Suppose, for example, that the morphogenetic substance is initially synthesized and degraded at a uniform rate by all cells within the slug. Assume also that the rate of its diffusion across the slime encasement to the exterior of the slug is inversely proportional to the thickness of the sheath. Then the concentration of the molecule will be higher in the posterior regions of the slug where the sheath is thick, and diffusional loss will be slow, while its concentration in the anterior region will be lower because the sheath is relatively thin, and diffusional loss will be more rapid (Figure 5.3). If this molecule promotes prespore differentiation and/or inhibits prestalk determination, the relative concentrations of the molecule within the slug can give rise to the two cell types. The position of a cell within the slug determines the extent of exposure to the morphogen; morphogen exposure promotes or inhibits expression of a new program of gene expression (reversible commitment); and the genetic program allows expression of a new complement of cell-type-specific proteins (biochemical differentiation). At some time during culmination cell destiny becomes determined, and at this time we say that the cell is "irreversibly committed." When position in a developing organism determines cell fate, we refer to the positional analysis as a *fate map*.

If morphogenic substances do, in fact, regulate cell determination in *Dictyostelium*, it should be possible to demonstrate an influence of chemical agents on the formation of prestalk and prespore cells. Relevant to this suggestion, it has been shown that cyclic AMP promotes differentiation of both cell types. More interestingly, however, ammonium ions are apparently required for spore formation, at least in certain mutant strains of *Dictyostelium*. In these strains, an increase in the ammonium ion concentration shifts the ratio of prespore to prestalk cells in favor of the former. Consequently, ammonia, generated during

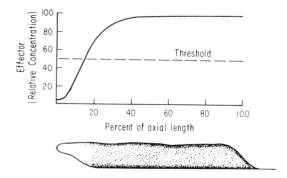

Figure 5.3. Proposed relationship between concentration of morphogen and position within a migrating slug. The morphogen is considered to be synthesized and degraded by the amoebae at uniform rates, so that the rate of diffusion through the sheath encasement determines the morphogen concentration. This concentration is lowest at the anterior tip of the slug because it is in this region that the sheath is thinnest, allowing maximal rates of loss of the morphogen by diffusion. From W.F. Loomis in *Developmental Biology, Pattern Formation and Gene Regulation*, ICN–UCLA Symposium on Molecular and Cellular Biology (D. McMahon and C.F. Fox, eds.), W.A. Benjamin, Inc., 1975, reprinted with permission.

early slug migration by the constituent amoebae, represents one candidate for a morphogenetic substance that could initiate differentiation of a single cell type into two morphogenetically and biochemically distinct cell types.

Heterocyst Development in Blue-Green Bacteria (*Anabaena*)

Vegetative, filamentous cyanobacteria such as *Nostoc* and *Anabaena* give rise to two types of differentiated structures, a hard-walled asexual spore called an *akinete,* and a nondividing nitrogen-fixing cell referred to as a *heterocyst.* Differentiation of a vegetative cell to an akinete is triggered by carbon starvation or by other environmental conditions unfavorable to growth, but differentiation to a heterocyst is specifically promoted by removal of ammonium and organic sources of nitrogen. Of all prokaryotic differentiation processes, heterocyst formation has been considered to resemble somatic cell development in animals most, first, because heterocyst development is subject to both temporal and spatial control, and second, because the process is essentially irreversible. A heterocyst cannot divide and apparently does not revert to vegetative growth, even upon addition of an organic nitrogen source. Thus, irreversible programs subject to both temporal and spatial regulation were probably available in the prokaryotic world before the advent of the eukaryotic cell.

Nitrogen fixation is a process found exclusively in prokaryotes, and the nitrogenase enzyme complex is active only in an anaerobic or microaerophilic environment. In bacteria capable of anaerobic growth, nitrogenase can exist within the vegetative cell, but in Cyanobacteria, which evolve oxygen photosynthetically, sensitivity of nitrogenase to oxygen poisoning necessitates compartmentalization—separation of the photosynthetic apparatus from the nitrogen-fixing machinery. Since the heterocyst is a multilayered thick-walled cell that probably exhibits low permeability to oxygen, it would be reasonable to assume that a pre-existing spore differentiation program was utilized and modified during evolution of the nitrogen-fixing compartment.

In order to ensure that co-existing photosynthetic vegetative cells and heterocysts could supply the other cell type with essential nutrients (sources of carbon and energy versus sources of organic nitrogen, respectively), *Anabaena* and other filamentous blue-green bacteria evolved a spatial program of heterocyst development that ensures the presence of a single heterocyst at regular intervals along a chain of vegetative cells. Figure 5.4 illustrates the process. The normal spacing between heterocysts in different species of *Anabaena* varies, but is about ten photosynthetic cells. After all of the vegetative cells divide, each heterocyst (which does not divide) is initially separated by about 20 cells. In the center of the chain, several cells begin to swell into heterocyst precursor cells called *proheterocysts.* Only one of these, however, develops into a heterocyst, and the other adjacent cells regress back to the vegetative state. This process restores

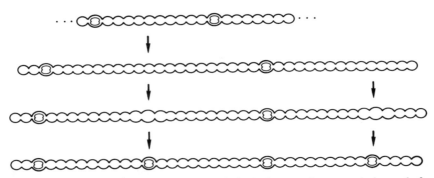

Figure 5.4. Schematic depiction of the morphological events that occur during and after division of vegetative cells in *Anabaena* in order to maintain approximately constant spacing between heterocysts.

the normal interval of ten vegetative cells per heterocyst (Figure 5.4). The mechanism responsible for the maintenance of this spatial pattern will be discussed below.

The Temporal Program for Heterocyst Formation

Differentiation of a vegetative cell into a heterocyst in response to nitrogen starvation involves a large number of morphological and biochemical changes. Many of these changes function to ensure anaerobiosis in the internal compartment of the heterocyst. From the relatively thin cell wall of the vegetative cell, a thick-walled structure, which presumably restricts oxygen exchange, must be constructed. As shown in Figure 5.5, the heterocyst envelope consists of a five-layered structure plus a membrane. These layers have been named on the basis of their appearance in the electron microscope. Adjacent to the cell membrane is a vegetative-cell-wall derived layer, then a laminate layer, the subhomogeneous and homogeneous layers, and finally, the outer fibrous layer and mucilagenous sheath. Clearly many biosynthetic activities not expressed in the vegetative cell are required for the production of this structure.

Nitrogen fixation has been estimated to require about 18 moles of ATP and several reducing equivalents per molecule of nitrogen gas reduced to ammonia. The heterocyst must therefore generate enormous quantities of ATP and NADPH. Yet this must occur in an oxygen-depleted medium. In order to prevent oxygen evolution during photosynthesis without complete loss of the photosynthetic ATP-generating capacity of the heterocyst, the functioning of photosystem II, responsible for oxygen generation, is blocked without loss of photosystem I or the photosynthetic electron transfer chain. This specific change allows *cyclic* phosphorylation of ADP to ATP to occur as shown in Figure 5.6.

In order to maintain a reducing environment, required both for nitrogen reduction and for the maintenance of active nitrogenase, the enzymes of the pentose

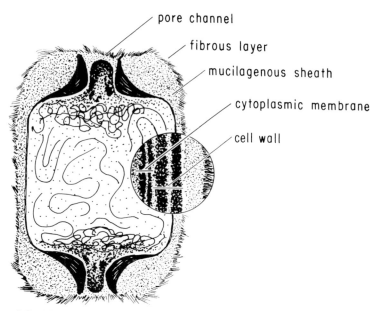

Figure 5.5. View of a heterocyst showing cell wall layers that were first revealed by electron microscopic studies.

phosphate pathway are induced to high levels. For example, levels of 6-phosphogluconate dehydrogenase and glyceraldehyde 3-phosphate dehydrogenase increase 20-fold. This allows for rapid production of NADPH from carbon sources provided by the dark reactions of photosynthesis, which occur in the vegetative cell. Electron carriers involved in nitrogen reduction (such as flavodoxin) are also induced to high levels. By contrast, the enzymes of the reductive pentose phosphate shunt and those responsible for carbon dioxide fixation (such as ribulose diphosphate carboxylase) are absent in the heterocyst. Consequently, reducing power can be generated within the heterocyst but is utilized only for nitrogen reduction and for a few essential cell functions.

Development of a proheterocyst into a mature heterocyst takes about 60 hr and has been arbitrarily divided into four stages (Figure 5.7A). As described in Chapter 4 for endosporulation in *Bacillus,* specific biochemical events occur during each of these stages. During stage I, the developing proheterocyst becomes partially detached from the vegetative filament, and the outer fibrous layer of the cell wall appears. During stage II the connecting bottle-necked channels or junctions between heterocysts and vegetative cells, the *microdesmata,* form. The homogeneous layers of the cell wall form, and the heterocyst becomes resistant to the action of lysozyme. It is during this stage that the nitrogenase protein is made, but it is present in an inactive state (Figure 5.7A). Stage III is accompanied by appearance of the laminate layer of the cell wall, and the nitrogenase enzyme, synthesized largely during stage II, is activated. Finally, morphological and

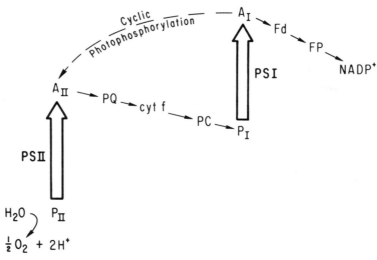

Figure 5.6. Schematic depiction of the photosynthetic process in blue-green bacterial vegetative cells. Photosystem II (PSII) catalyzes the photon-dependent activation of an electron derived from water. Oxygen is evolved in this process, and electrons are passed from a pigment (PII) to an electron acceptor (AII). Subsequently, the electron passes down the photosynthetic electron transfer chain to components including plastoquinone (PQ), cytochrome f (cyt f) and plastocyanin (PC) in a process that is coupled to ATP synthesis. Photosystem I uses a second photon to further activate the electron from a pigment, PI, to a higher energy state when the electron is transferred to a new acceptor, AI. Blockage of photosystem II in the heterocyst prevents oxygen evolution but permits ATP synthesis by the cyclic phosphorylation process involving photosystem I and the electron transfer chain as shown by the dashed line.

functional development of the proheterocyst is completed during stage IV, and the mature heterocyst emerges. It is during this stage that heterocyst development becomes fully irreversible. Addition of organic nitrogen to the heterocyst-containing filament does not result in reversion of the heterocyst to a photosynthetic vegetative cell.

Throughout this temporal program, vegetative proteins are extensively degraded, and protein turnover occurs rapidly. As depicted in Figure 5.7B, some proteins are (I) synthesized continuously, before, during and after proheterocyst development while (II) other vegetative proteins are made only up to a specific time during development. A third class of proteins (III) is made transiently during development, while a fourth group of proteins (IV) are first made during proheterocyst development and continue to be made in the mature heterocyst. As also demonstrated for sporulation in *Bacillus,* different patterns of gene expression are thus observed during heterocyst development. Gene expression clearly involves temporally controlled genetic activation and deactivation for transcription and translation, and it is therefore not surprising that chloramphenicol (an inhibitor of bacterial protein synthesis) blocks development at all stages.

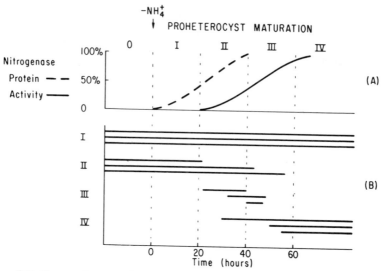

Figure 5.7. Temporal pattern for synthesis and activation of nitrogenase (A) as well as for the syntheses of various proteins detected by gel electrophoretic techniques (B). These proteins fall into the four classes shown in (B): (I) those synthesized continuously; (II) those synthesized continuously up to but not past some stage during heterocyst development; (III) those synthesized only for a short period during proheterocyst differentiation but not before or after this period, and (IV) those synthesized in the mature heterocyst, but not prior to a specific time during development.

Spatial Regulation of Heterocyst Development

As discussed previously, heterocyst development is subject to rigid spatial regulation so that one can predict with fair reliability where new heterocysts will appear. For example, in *Anabaena cylindrica* the spacing is 9.2 ± 2.8 vegetative cells per heterocyst while in *A. catena* this spacing is 10.1 ± 2.5 cells per heterocyst. Development of an internal proheterocyst, flanked by filaments containing heterocysts on both sides, occurs with highest frequency at a position 50% along the internal length of the vegetative filament (Figure 5.8). By contrast, when the developing proheterocyst is in a terminal position on the chain with a heterocyst on only one side, it occurs with highest frequency 70% of the length from the neighboring heterocyst to the end of the cell chain. In both cases, the curves are tighter than would have been predicted from a normal Gaussian distribution. These observations have led to two hypotheses regarding heterocyst positioning: (1) development of proheterocysts is controlled by an inhibitor of proheterocyst development produced by the heterocyst itself and by proheterocysts that have developed beyond a certain stage; and (2) *unequal mitosis* is at least partially responsible for deviation from a Gaussian distribution of heterocysts along the chain.

Unequal mitosis refers to the fact that in blue-green bacteria, as in the normal

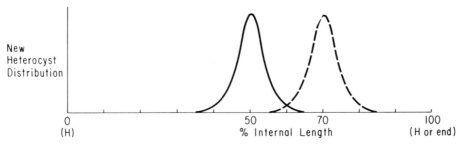

Figure 5.8. Distribution of heterocysts along filaments of vegetative *Anabaena* cells when the developing proheterocysts are in the middle of a chain of greater than ten cells (—) or in a terminal position (– – – –) with a heterocyst only on one side. In the former case, the distribution of heterocysts centers around the middle of the chain, while in the latter case the new heterocyst develops with greatest probability 70% from the heterocyst and 30% from the chain terminus. These results can be explained by the inhibitory morphogen hypothesis if it is assumed that the ends of the cell filament are closed so that morphogen is reflected back along the chain (see Fig. 5.10). Both curves are tighter than a normal Gaussian distribution. H = heterocyst.

cell cycle of *Caulobacter* and the terminal DNA replicative event in *Bacillus* prior to sporulation, cell division yields two daughter cells of unequal size. Observations of cell division in filamentous strains of *Anabaena* have revealed that unequal mitosis occurs with positional regularity. Thus, if a cell has arisen as the left daughter cell, its left daughter will be smaller, but if the cell has arisen as the right daughter cell, its right daughter will be smaller. This rule is illustrated in Figure 5.9. The second observation relates to heterocyst development. The heterocyst always arises from a smaller daughter cell if unperturbed. These observations, while not understood at a mechanistic level, provide an explanation for deviation of heterocyst appearance in a filamentous chain of vegetative cells from a Gaussian distribution.

In order to explain the positioning of new heterocysts 50% along an internal filament, and 70% along a terminal filament (Figure 5.8), it has been proposed that heterocysts and proheterocysts produce (and develop resistance to) an inhibitor of heterocyst development. This inhibitor is presumed to be a small substance (morphogen) that can diffuse along the chain from one cell to another without rapid loss to the external medium. It is either degraded or lost from the chain at a uniform rate, and its production is minimal in vegetative cells. Consequently, the distribution of the inhibitor is as shown in Figure 5.10. Its concentration should be lowest at the midpoint between two heterocysts in an internal chain, and 70% along the chain for development within a filament flanked only on one side by a heterocyst. Thus, the experimental observations regarding positioning of heterocyst development can be explained by the inhibitory morphogen theory.

Further support for the inhibitor idea stems from microscopic observational studies of heterocyst development following vegetative cell division. Upon ex-

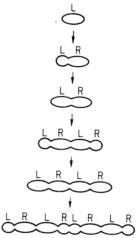

Figure 5.9. Pattern of cell division in *Anabaena*. L refers to a cell produced on the left-hand side; R refers to one produced on the right. Each cell division event gives rise to daughter cells of unequal size and competence for heterocyst development. Only the smaller daughter cell can normally develop into a heterocyst.

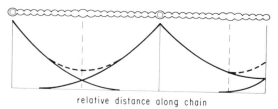

relative distance along chain

Figure 5.10. Distribution of the postulated inhibitor of heterocyst morphogenesis in filaments of *Anabaena* grown in the absence of ammonium ions and organic nitrogen. *Left:* Distribution along an internal filament of vegetative cells flanked by heterocysts. *Right:* Distribution along a terminal chain of vegetative cells in which a heterocyst is found only on one side. It is assumed that a terminal cell provides a barrier to inhibitor flow causing reflection and a consequent backwards flux of the inhibitor. The proheterocysts develop most rapidly at the point along the chain where the inhibitor concentration is lowest.

tension of the chain by mitosis, several cells, appropriately positioned, swell as if progressing toward proheterocyst development as shown in Figure 5.4. Subsequently, one of these cells continues to develop while the remainder regress back to vegetative growth. This behavior can be explained by assuming that at the time regression of adjacent cells occurs, the dominant (most highly developed) proheterocyst has begun to secrete an inhibitor of heterocyst development and has simultaneously developed resistance to its action.

Studies have been carried out to determine when heterocyst development becomes irreversible. These studies have led to the suggestion that commitment in Cyanobacteria may involve a probability function, and that proheterocysts become totally incapable of regression only during stage IV (Figure 5.7). Interestingly, an amino acid analogue, azatryptophan, decreases the probability of regression during early stages of development. This compound also reduces the spacing of heterocysts along the vegetative cell filament at micromolar concentrations. A reasonable explanation would be that azatryptophan reduces the concentration of active inhibitor. Possibly, the inhibitor is a tryptophan-containing peptide, and synthesis in the presence of azatryptophan either slows the biosynthetic process or gives rise to a less potent inhibitory peptide containing azatryptophan. The azatryptophan effect therefore provides support for the inhibitor hypothesis. Because this compound also decreases the time during development when heterocyst differentiation appears irreversible, it can be concluded that the inhibitor concentration influences both the time of apparent commitment and heterocyst spacing.

Control of Morphogenesis in *Hydra*

Hydra is a small freshwater coelenterate, about 5 mm long, consisting of about 100,000 cells of about 12 cell types. The structure of the organism is shown in Figure 5.11A. There are two principal cell layers—an external ectoderm and an internal endoderm, separated by a primitive nerve net, the *mesoglia* (Figure 5.11B). A live food source is caught and paralyzed by means of sting cells present in the tentacles in the head region, and the prey is subsequently passed through the mouth or *hypostome* into the gastric cavity, bounded by the *gastric column,* where it is digested and ingested. The *basal disc* anchors the organism to the substrate. The organism reproduces asexually by budding from the external wall of the gastric column at a position about one third of the way from the basal disc (Figure 5.12). After reaching a certain size, the bud becomes autonomous. *Hydra* is the most complex of the differentiating organisms for which cell dissociation and complete organismal regeneration is possible.

A variety of regeneration experiments have led to the conclusion that morphogenesis in *Hydra* is controlled by small, diffusible morphogenetic hormones or morphogens. According to one well-substantiated postulate, there may be two such morphogens controlling head and bud formation: a *head activator* and an inhibitor of head morphogenesis. The inhibitor is thought to be made by nerve cells in the head region and to diffuse from the head down the gastric column toward the basal disc. By contrast, the head activator may be made throughout the gastric column. This creates unequal gradients of the two substrates as shown in Figure 5.13. It is presumably the *concentration* of the inhibitor that determines bud position and the *ratio* of head activator to inhibitor that reinforces head formation. Bud formation occurs at the position of minimal inhibitor concentration only when this concentration is below some threshold value.

Early studies suggested that both the activator and inhibitor might be small peptide-hormone-like substances, effective at exceptionally low concentrations

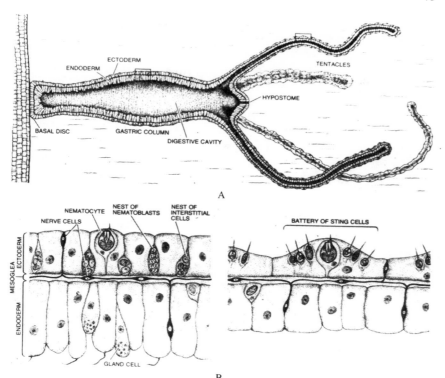

and present in graded concentrations throughout the organism. These substances are probably maintained in an inactive state within vesicles in nerve cells in the head region before release into the mesoglia. The vesicles are probably analogous to neurosecretory vesicles in higher animals.

Recently the head activator has been isolated from two closely related coelenterates: sea anemones and *Hydra*. In the sea anemone, 1 kg of material yielded about 0.1 μg of the purified material, obtained with a 20% yield. Purification was in excess of a billionfold. Thus, 1 kg of organisms contained only 0.5 μg

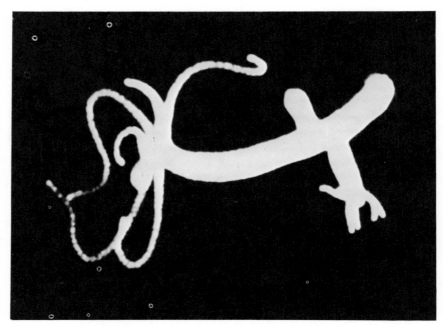

Figure 5.12. Asexual reproduction in *Hydra* by budding. While individual cells divide every 24 hr, optimal growth results in budding every 3–4 days. The first bud appears along the gastric column about two–thirds of the way from the head to the basal disc. When growth is rapid, a second bud may appear opposite the first, but somewhat closer to the head. The positions of these buds are consistent with a control mechanism involving diffusible morphogens made in the head and bud regions. From Hydra as a model for the development of biological form, A. Gierer. Copyright © 1974 by Scientific American, Inc. All rights reserved.

of head activator. Similar amounts were obtained from *Hydra*. The structure of the head activator from these two sources was found to be identical:

$$\text{p-glu-(pro)}_2\text{-(gly)}_2\text{-ser-lys-val-ile-leu-phe}$$

an 11-amino-acid peptide with a pyroglutamyl residue at the N-terminus and a phenylalanyl residue at the C-terminus. The C-terminus is hydrophobic while the N-terminus is moderately hydrophilic, and a single charged amino acid (lysine) occurs in the center.

The head activator has been shown to initiate head-specific growth and differentiation without stimulating growth of the basal disc. It appears to trigger cells to divide and promotes the differentiation of multipotent stem cells (interstitial cells) into nerve cells. While the detailed mechanism of its action remains to be elucidated, this compound represents the first known morphogen of defined structure that influences the spatial distribution of cells within an organism. It would not be surprising if similar peptides function in the spatial control of

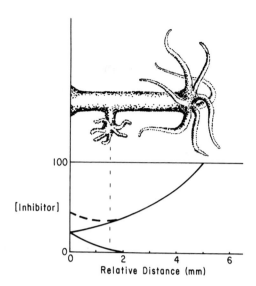

Figure 5.13. Proposed gradient of the head inhibitor along the gastric column.

morphogenesis of other simple and complex organisms, from the prokaryotic blue-green bacteria to the most complex eukaryotes. It is also of interest to note that peptide pheromones (i.e., in yeast) and hormones (in higher organisms) influence development and sexual activities in these organisms (see Chapter 9).

Selected References

Adams, D.G. and N.G. Carr. "The Developmental Biology of Heterocyst and Akinete Formation in Cyanobacteria" in *CRC Crit. Rev. Micro. 9:*45 (1981).

American Society of Zoologists. *Principles and Problems of Pattern Formation in Animals,* Symposium of the Division of Developmental Biology, 1980, *American Zoologist,* Vol. 22, (1982).

Ashworth, J.M. and J.E. Smith, eds. *Microbial Differentiation,* Society for General Microbiology Symposium 23, Cambridge University Press, 1973.

Bode, H.R. and C.N. David. Regulation of a multipotent stem cell, the interstitial cell of *Hydra, Prog. Biophys. Molec. Biol. 33:*189 (1978).

Bonner, J. T. Differentiation in social amoebae, *Scient. Am. 201:*152 (1959).

Fleming, H. and R. Haselkorn. The program of protein synthesis during heterocyst differentiation in nitrogen-fixing blue-green algae, *Cell 3:*159 (1974).

Gierer, A. Hydra as a model for the development of biological form, *Scient. Am. 231:*44–59 (1974).

Lenhoff, H.M. Behavior, hormones, and *Hydra, Science 161:*434 (1968).

Loomis, W.F. *Dictyostelium discoideum, A Developmental System,* Academic Press, New York, 1975.

Loomis, W.F., "Genetic analysis of development in *Dictyostelium*" in *The Molecular Genetics of Development* (T. Leighton and W.F. Loomis, eds.), Academic Press, New York, 1980.

Loomis, W.F., ed. *Development of Dictyostelium,* Academic Press, New York, 1982.

Morrissey, J.H. "Cell proportioning and pattern formation," in *The Development of Dictyostelium discoideum* (W.F. Loomis, ed.), Academic Press, San Diego, 1982.

Schaller, H.C. and H. Bodenmüller. Isolation and amino acid sequence of a morphogenetic peptide from *Hydra, Proc. Natl. Acad. Sci. USA 78:*7000 (1981).

Wilcox, M., C.J. Mitchison, and R.J. Smith. Pattern formation in the blue-green alga, *Anabaena.* I. Basic mechanisms, *J. Cell Sci. 12:*707 (1973).

Wilcox, M., G.J. Mitchison, and R.J. Smith. Pattern formation in the blue-green alga *Anabaena.* II. Controlled proheterocyst regression, *J. Cell Sci. 13:*637 (1973).

CHAPTER 6

Transmembrane Transport and the Control of Bioelectrical Activities

> The water continually flowed and flowed
> and yet it was always there;
> it was always the same and yet
> every moment it was new.
>
> *Hermann Hesse*

In the preceding chapters, we considered several examples of developmental events that occur in microorganisms. In many instances we saw that intercellular communication as well as the ability of individual cells to sense changes in their environment were necessary for these processes to occur. External stimuli of various sorts were able to effect changes in the structure, metabolism and biochemistry of differentiating cells. In all cases so far examined, these sensory detection systems include proteins in the cell membrane whose roles are to transduce information from the outside of the cell into signals that then modulate various intracellular biochemical events. Some of these proteins are those also responsible for the transmembrane transport of solute molecules, while others appear to be more highly specialized for their chemoreceptor function. In later chapters, we shall explore the molecular details of some of these communication mechanisms. As a foundation for this, however, we shall first consider some mechanisms by which many cells directly communicate with their environment via transmembrane transport.

The transport of solute molecules across cell membranes serves all living cells for the acquisition of nutrients, the maintenance of a cytoplasmic ionic composition consistent with life, and the generation of bioelectric signals. A few solute permease systems chemically modify the substrates they transport, thereby initiating sequences of catabolic or anabolic reactions. These permeases may also regulate the cytoplasmic concentrations of regulatory molecules that control rates of transcription and/or translation. Some of these regulatory molecules are actually acted upon and transported by the permeases while the cytoplasmic concentrations of others are indirectly controlled. In still other cases, the transport systems regulate the states or cytoplasmic concentrations of *second messenger proteins*, which either regulate other cellular physiological processes or act directly on the protein biosynthetic machinery. Finally, some transport proteins have gained receptor functions that provide signaling mechanisms, conveying information to cellular organelles responsible for motility and behavior.

These facts, which have been established only in recent years, have led to the recognition of probable evolutionary relationships between metabolic enzymes, transport proteins (this chapter), cell surface ligand receptors (Chapter 8), macromolecular receptors (Chapter 7), and biochemically and bioelectrically mediated regulatory signaling devices (this chapter and Chapters 8 and 9). In fact, the occurrence of present-day multifunctional transmembrane proteins, which serve as permeases, metabolic enzymes, chemoreceptors, components of second messenger relay systems and transcriptional regulatory agents, shows that these evolutionary relationships are not just probable, they seem to be well established. This fact, in turn, leads to the likelihood that unifunctional enzymes, permeases, receptors, and regulatory proteins all evolved from common ancestral proteins. In this and the next chapter, well-characterized bacterial transport and regulatory systems will be discussed in some detail. Their interrelationships will be emphasized, particularly when multifunctionality leads to evolutionary insights. Because the early morphogenic and developmental processes had to result from expansion of, or superimposition of, new regulatory mechanisms upon the preexisting control processes (Chapters 4 and 5), the existence of mechanistically similar processes in the less well-understood differentiating microorganisms and multicellular organisms can be anticipated. These processes therefore serve as a foundation for a molecular understanding of the regulatory constraints imposed on the most complex forms of multicellularity.

Classification of Transport Systems in Bacteria

Bacteria have evolved a variety of mechanisms by which solutes are transported into and out of living cells, and very similar mechanisms are operative in eukaryotes. In virtually all living cells, H^+, Na^+, and Ca^{2+} are actively extruded, while K^+ and Mg^{2+} are accumulated within the cytoplasm against substantial concentration gradients. These ion transport processes occur as a result of the expenditure of a *primary* source of metabolic energy or by the utilization of *secondary,* pre-existing ion gradients. For example, in glycolyzing *E. coli* cells, protons are actively extruded while K^+ is accumulated in processes catalyzed by proton translocating and K^+ translocating ATPases that couple ion transport to the hydrolysis of ATP. In these same cells, Na^+ and Ca^{2+} are extruded by two separate *antiport* systems which exchange the cation for H^+ and thus utilize the proton electrochemical gradient as the energy source that drives the active expulsion of Na^+ or Ca^{2+} from the cytoplasm.

Nutrients such as carbohydrates, amino acids, vitamins, and anions, are transported by a variety of mechanisms. For example, five protein-mediated carbohydrate transport mechanisms are known to occur in microorganisms. These are represented schematically in Figure 6.1, where a specific example of each type of system found in *E. coli* is depicted. These systems can be classified conve-

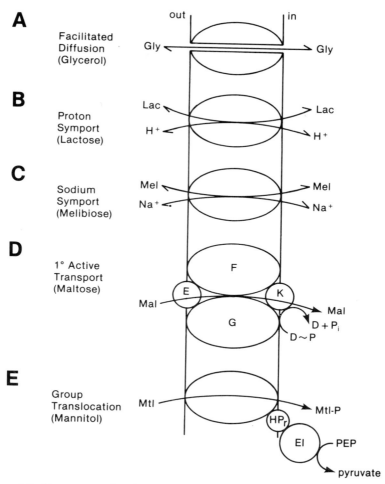

Figure 6.1. Permease-mediated transmembrane carbohydrate transport processes that occur in *E. coli*. Glycerol (Gly) crosses the membrane by passive diffusion through a nonstereospecific protein pore. Lactose (Lac) and melibiose (Mel) enter the cell by cation symport involving protons and sodium ions, respectively. Each of these three transport processes is probably catalyzed by a single protein species encoded by a specific gene on the bacterial chromosome. Maltose (Mal) transport may involve the hydrolysis of a high-energy phosphoryl donor (D $\sim$ P) that drives the active accumulation of the sugar. Four distinct proteins, encoded by four different genes within the *mal* regulon, are required for transport. Finally, the transport of mannitol (Mtl) involves three proteins of the phosphotransferase system, the general energy-coupling proteins of the system, Enzyme I (EI) and HPr, as well as the integral Enzyme II specific for mannitol (EII^{Mtl}). Each of these proteins is sequentially phosphorylated at the expense of phosphoenolpyruvate (PEP) before translocation of mannitol and its coupled phosphorylation can occur.

niently according to the source of energy coupled to the transport process. Facilitated diffusion is a process not coupled to metabolic energy and is, therefore, not capable of accumulating a substrate against a concentration gradient. Glycerol crosses bacterial cell membranes by this mechanism. Other systems are grouped together as *active* or *concentrative* transport systems since they are energy coupled and accumulate substrates against concentration gradients. The ability of an active transport system to concentrate its substrates on one side of a membrane must involve alteration of the *carrier* while the solute is passing from one side of the membrane to the other. The force driving this alteration is derived from specific products of cellular metabolism, and it is this source of metabolic energy that is used to classify the active transport systems. Examples of these types of systems occurring in *E. coli* include the lactose permease, which utilizes a proton electrochemical gradient as the source of energy for solute accumulation, the melibiose permease, which functions by Na^+:sugar cotransport, and the maltose transport system, which is thought to utilize a chemical form of energy to concentrate its substrates. Finally, glucose and mannitol are substrates of the phosphotransferase system (PTS), and in the process catalyzed by this system, the substrate is phosphorylated during transport. Such a mechanism is referred to as group translocation. In the following sections we shall discuss these different transport processes, exemplifying each of these with the best characterized permeases in *E. coli*.

Solute Facilitators

Solute facilitators equilibrate their substrates across the membrane without accumulating them against net concentration gradients. Because the process is not coupled to a primary source of metabolic energy, it is conceptually and possibly mechanistically very simple. In nature two types of facilitators are found, those behaving as nonspecific transmembrane pores, and those exhibiting a high degree of stereospecificity for their substrate. The former systems allow passage of any solute of the appropriate size and charge specifications. The glycerol facilitator in *E. coli* is such a protein. The internal pore size of this permease is about 0.4 nm, and it transports any straight chain, neutral carbohydrate that fits easily into the pore. As expected, transport activity is fairly temperature insensitive and nonsaturable with respect to substrate concentration. A single type of protein is probably involved.

In *E. coli* there are no known nonconcentrative but stereospecific facilitators, although such transport systems are found in other cells. There *are* chemiosmotically coupled facilitators that couple solute transport to the transport of specific ions, however. The lactose and melibiose permeases represent stereospecific facilitators that couple transport of their substrate sugars to the translocation of a cation, H^+ and Na^+, respectively, for these two systems. Except for the sugar and cation specificities of the two systems, they probably function by very similar mechanisms.

The lactose permease has been purified, and its transport function has been reconstituted in artificial phospholipid membranes (proteoliposomes). It is therefore certain that a single protein (of known amino acid sequence) catalyzes the coupled transport of one molecule each of lactose and H^+. Transport of just the sugar or the proton cannot occur without the other species being carried across the membrane. Considerable evidence suggests that the imposition of a transmembrane electrical potential changes the conformation of the permease protein and thereby alters its catalytic transport characteristics.

The lactose and melibiose permeases are highly stereospecific for their sugar substrates and also exhibit a high degree of cation specificity. In these respects they resemble most metabolic enzymes. They therefore must possess recognition sites for sugar and cation. Possibly the two species bind to the carrier on one side of the membrane, and a conformational change in the protein exposes them to the opposite side. The details of such a translocation process are not known.

Chemically Driven Active Transport

Maltose is transported into the *E. coli* cell by a complex system consisting of four distinct proteins, all of which are essential for transport activity (Figure 6.1). One of these proteins is water soluble and binds maltose with high affinity. Consequently, it is called the maltose-binding protein. Because it is encoded by the *malE* gene, it is sometimes called the E protein. It is localized to the external surface of the cytoplasmic membrane where it can associate (in the presence of maltose) with the transmembrane protein constituents of the transport system. This protein also associates with a transmembrane constituent of the bacterial chemoreception system, thus serving as a chemoreceptor (see Chapter 8).

Two of the remaining three proteins comprising the maltose permease are integral constituents of the membrane. These two proteins are encoded by the *malF* and *malG* genes and are therefore referred to as the F and G proteins. It is thought that they may form a stereospecific pore through which maltose can pass. At least one maltose-binding site within this pore region of the permease has been identified, and this site confers stereospecificity on the system, even in the absence of the maltose-binding protein. Association of the maltose–maltose binding protein complex with the external surface of the integral permease constituents, however, appears to induce a conformation of the entire complex that is favorable to transport. Finally, a fourth protein, encoded by the *malK* gene and called the K protein, is localized to the cytoplasmic surface of the permease complex, being specifically and tightly bound to the integral membrane proteins of the system. There is some evidence that this protein functions in the allosteric regulation of maltose transport, and it may also be involved in the transcriptional regulation of the genes coding for the maltose transport and catabolic enzyme systems. Since all four of the protein constituents of the maltose permease are essential for transport activity, the two water soluble E and K proteins, localized to the external and internal surfaces of the membrane, respectively, may be bifunctional; the maltose-binding protein functions in transport and chemoreception, while the cytoplasmic K protein may function in transport and regulation. The involvement of transport proteins in transcriptional and metabolic regulation will be discussed in the next chapter.

Finally, the maltose transport system is classified as a primary active transport system because it apparently utilizes chemical or electrical energy in order to translocate maltose across the membrane and accumulate it in the cytoplasm.

Group Translocation Catalyzed by the Bacterial Phosphotransferase System

The phosphotransferase system (PTS) in *Escherichia coli* catalyzes the transport of many sugars including glucose, fructose, and mannitol. Figure 6.2 indicates the sequence of phosphoryl transfer reactions that is essential for the process of sugar group translocation. In this sequence, a high-energy phosphoryl group is transferred first from phosphoenolpyruvate (the phosphoryl donor) to Enzyme I and then to a small heat stable phosphoryl transfer enzyme called HPr. These two proteins are general cytoplasmic constituents of the PTS that are required for the group translocation of all sugar substrates of the system. The phosphoryl transfer reactions catalyzed by these two enzymes involve the intermediate phosphorylation of histidyl residues in the proteins.

Remaining proteins of the PTS exhibit sugar specificity. For example, glucose transport involves two proteins, a soluble Enzyme III^{glc} of 20,000 molecular weight, phosphorylated at the expense of phospho HPr, and an integral membrane Enzyme II^{glc} of about 50,000 molecular weight which is phosphorylated by Enzyme III^{glc}. Phospho Enzyme II^{glc} then donates its phosphoryl group to glucose, which is concomitantly translocated and phosphorylated. Glucose-6-phosphate is released intracellularly. On the other hand, if mannitol is present in the external medium, the phosphoryl group of HPr is transferred to Enzyme II^{mtl}, an integral, transmembrane protein of 68,000 molecular weight, and this phosphorylated enzyme effects the translocation of the sugar across the membrane, releasing mannitol-1-phosphate in the cytoplasm. No soluble sugar-specific Enzyme III is involved in mannitol uptake. Mannitol-1-phosphate is oxidized to fructose-6-phosphate in a reaction in which NAD^+ is reduced to NADH. This reaction is catalyzed by the mannitol-1-phosphate dehydrogenase, a cytoplasmic protein of 40,000 molecular weight. Fructose-6-phosphate feeds directly into the glycolytic pathway that generates phosphoenolpyruvate. Thus, the end product of the metabolism of a sugar substrate of the PTS is the energy source that drives uptake of the sugar.

In addition to glucose and mannitol, the PTS translocates many other sugars across the membrane. These sugars include fructose, mannose, N-acetylglucosamine, glucitol and galactitol among others, and for each of these sugars there

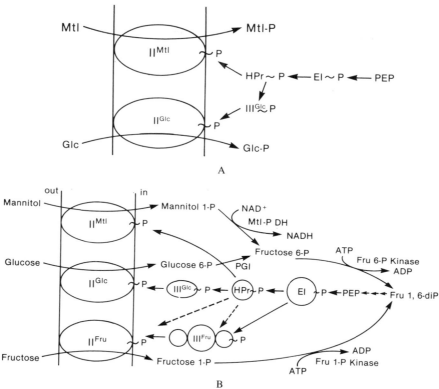

Figure 6.2. (A) Schematic depiction of the phosphoryl transfer chain of the bacterial phosphotransferase system showing the enzyme constituents responsible for the transport and phosphorylation of mannitol (Mtl) and glucose (Glc). The abbreviations are as follows: PEP, phosphoenolpyruvate; EI, Enzyme I; HPr, heat stable phosphocarrier protein of the PTS; EIII^{Glc}, the glucose-specific Enzyme III of the PTS; EII^{Glc} and EII^{Mtl}, the glucose-specific and the mannitol-specific Enzymes II of the PTS, respectively. The Enzymes II are integral membrane proteins that function as the sugar permeases. The remaining enzymes depicted are soluble, or peripherally associated with the membrane. Each protein must be phosphorylated in sequence in order for group translocation of the sugar to occur. (B) Integration of the phosphotransferase system into the glycolytic reactions to form a "glycolytic cycle" in which PEP produced in glycolysis, is used for sugar uptake.

exists in the membrane a sugar-specific Enzyme II. Not only do these proteins transport and phosphorylate their sugar substrates, they also serve as the chemo-receptors that transmit information to the basal region of the flagellum, thereby directing the motile behavior of the organism (Chapter 8). Finally, there is evidence that at least some of the Enzymes II regulate expression of the genes that code for the proteins of the PTS, and they also function catalytically to regulate the cyclic AMP biosynthetic enzyme, adenylate cyclase, as well as a variety of nonPTS sugar permeases, such as those for glycerol, lactose, melibiose, and maltose. The regulatory functions of the PTS will be discussed in the next chapter.

The multifunctionality of the PTS enzymes is therefore evident: Each Enzyme II is a sugar-specific chemoreceptor, permease, phosphoenolpyruvate-dependent kinase, and regulatory protein. The economy resulting from the use of a single protein in several related capacities must have provided some of the evolutionary pressure for the appearance of complex enzyme systems such as the phosphotransferase system.

One of the Enzymes II—that for mannitol—has been studied in considerable detail. Solubilization from the membrane and purification yields a single protein of 68,000 molecular weight. Because the gene that encodes this enzyme has been cloned and sequenced, its complete amino acid sequence is known. This protein, when reconstituted in a phospholipid bilayer, catalyzes mannitol transport in the absence of other membrane proteins. Moreover, two distinct group translocating reactions are catalyzed by this and all other Enzymes II (Figure 6.3). One of these involves the unidirectional transport of mannitol into the cell with its concomitant phosphorylation at the expense of phosphoenolpyruvate as discussed before (Figure 6.3A). The other involves a bidirectional exchange transport process in which the phosphoryl moiety of cytoplasmic mannitol phosphate is transferred to an incoming mannitol molecule. The sugar moiety of the cytoplasmic mannitol phosphate is expelled from the cell in this process. If the Enzyme II is transiently phosphorylated during this process, the bidirectional process can be thought of as a half-reaction of the overall, two-step, unidirectional transport process. Thus, unidirectional transport involves (A) phosphorylation of Enzyme II by phospho HPr followed by (B) transfer of the phosphoryl group from the Enzyme II to sugar. The exchange process may involve reaction (B) only, first in the reverse direction, with the transfer of phosphate from sugar

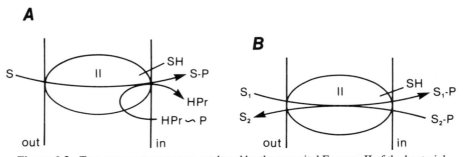

Figure 6.3. Two transport processes catalyzed by the mannitol Enzyme II of the bacterial phosphotransferase system. (A) Active group translocation involving phosphoryl transfer from phospho HPr to sugar. This transport process occurs in a unidirectional fashion. (B) Exchange group translocation involving phosphoryl transfer from a cytoplasmic sugar phosphate to sugar. Because the sugar moiety of the cytoplasmic sugar phosphate is expelled from the cell while the incoming sugar is taken up, this process is bidirectional. Abbreviations: S, sugar; S-P, sugar phosphate; II, Enzyme II of the PTS.

unidirectional

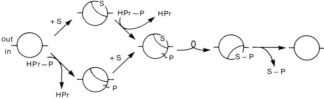

bidirectional

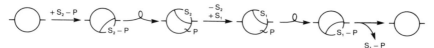

Figure 6.4. Simple mechanisms proposed for unidirectional and bidirectional transport of sugar catalyzed by an Enzyme II of the bacterial phosphotransferase system. This proposal suggests the involvement of a single protomeric enzyme that can exist in alternative phospho and dephospho forms. A more complicated model suggests the preferential catalysis of bidirectional transport by a dimeric Enzyme II complex.

phosphate to Enzyme II, and then in the forward direction with the transfer of phosphate to a different sugar molecule. The sugar phosphorylation process must occur in a vectorial fashion, where transfer of phosphate from enzyme to sugar is accompanied by transport in the inward direction while transfer of phosphate from sugar-phosphate to enzyme is coupled to sugar transport with outwardly directed polarity. The simplest possible mechanisms for the uni- and bidirectional processes involving a single Enzyme II protein are illustrated in Figure 6.4.

Some evidence suggests that the simple model depicted for bidirectional transport in Figure 6.4 is only an approximation to the true situation, which may be more complicated. Possibly two enzyme molecules form a dimeric complex in the membrane, and the two vectorial processes, with inward and outward polarity, may occur simultaneously and be preferentially coupled. In this model a single subunit (protomer) would possess an external sugar-binding site when phosphorylated, while the other dephosphorylated protomer would possess an internal sugar-phosphate-binding site. A single symmetrical molecular rearrangement would result in simultaneous transport of the sugars in opposite directions. While considerable evidence favors such a mechanism, more work will be required to establish that these concepts are correct.

Regulation of Transport Function by the Membrane Potential and the Control of Bioelectric Activity

An unequal distribution of ions across a biological membrane can give rise to a transmembrane electrical potential. Virtually all living cells maintain a membrane potential that is negative inside, partly because positively charged ions (H^+,

Na^+, and Ca^{2+}) are preferentially pumped out of the cytoplasm into the external medium. Ion-selective channels, then, can allow the passive movement of these ionic species through the membrane, down their concentration gradients, and ion transport can alter the magnitude of the membrane potential. The Na^+ channels of nerve cells selectively allow the passage of Na^+ down its electrochemical gradient from the external medium into the cell, while K^+ channels in the nerve membrane selectively allow passage of this ion down its electrochemical gradient, from regions of high concentration (inside the cell) to regions of lower concentration (outside the cell). Appropriate regulation of the activities of these two ion channel complexes can give rise to a sudden, transient change in membrane potential called an action potential. Membranes exhibiting the capacity to generate action potentials are said to be excitable.

In the case of the nerve cell, the electrical potential ($\Delta\Psi$) across the resting cell membrane is about 60 mV (interior negative). The arrival of an electrical or chemical stimulus of sufficient magnitude opens Na^+ channel proteins in the membrane, causing Na^+ influx into the cell near the point of stimulation. This local depolarization of the membrane continues for about 1 msec until $\Delta\Psi$ reaches a value of about 40 mV (interior *positive*). By this time, however, K^+ channel proteins have also been opened as a result of the change in $\Delta\Psi$, and K^+ flows out of the cell (Figure 6.5). Concomitantly, the Na^+ channels become inactivated, and $\Delta\Psi$ eventually returns to its resting value as shown in Figure 6.5. This action potential is conducted down the axon of the nerve cell as a wave of these

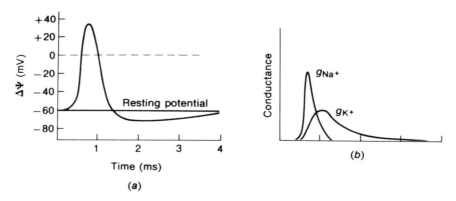

Figure 6.5. Mechanism of nerve impulse transmission. (a) A typical action potential that might be recorded at a single point along a nerve-cell membrane as a function of time. $\Delta\Psi$ is the transmembrane electrical potential (interior relative to exterior). (b) Changes in relative conductance (g) to Na^+ and K^+ as a function of time during the action potential shown in (a). A transient increase in g_{Na^+} leads to depolarization, while repolarization results from a subsequent temporary increase in g_{K^+}. These changes are partly explained by the sensitivities of the conformations of the ion channel proteins themselves to $\Delta\Psi$. From Zubay, BIOCHEMISTRY © 1983. Addison-Wesley, Reading, MA. Figure 30-4. Reprinted with permission. (Adapted from A.L. Hodgkin and A.F. Huxley, *J. Physiol.* *117*:500–544 (1952).)

depolarization–repolarization events. It may then be transmitted to other cells directly or through the release of certain compounds at the axon terminus called neurotransmitters.

The interrelationship of factors affecting the conformational states of nerve-cell ion channels (open and inactive) is complex and still not completely understood. However, it is clear that $\Delta\Psi$ is an important regulator of the activities of these proteins in the nerve-cell membrane. Thus, the entity that is controlled by the activities of the ion channels (the transmembrane electrical potential) also regulates their activities. It is this property of these proteins that leads to the excitability of sensory cell membranes.

It was once thought that electrical potential responsivity of ion channel proteins was a unique feature of nerve and muscle membranes, restricted to these tissues. We now know that numerous tissues exhibit the characteristics of excitability. In fact, single *pore* proteins in bacteria as well as solute transport proteins such as the lactose permease and several of the Enzymes II of the phosphotransferase system have been shown to be responsive to the membrane potential. In some cases, these responses may be of no physiological significance, but in other cases, regulation of transport by the membrane potential may allow the permease to sense the energy state of the cell and thereby transport an amount of nutrient consistent with the needs of the cell. Generally speaking, in bacteria, the greater the magnitude of the electrical potential across the cytoplasmic membrane, the higher is the energy state of the cell. Responsiveness of a transport system to the transmembrane electrical potential may be an inherent characteristic of the permease with no physiological relevance, or the phenomenon may have evolved great significance with respect to the control of cellular physiological processes.

Theoretical considerations lead to the conclusion that if a transmembrane protein is asymmetrically distributed across a membrane such that the charge distribution at one membrane surface differs from that at the other, the protein will possess a dipole moment. If this dipole moment is not perpendicular to the plane of the membrane or if its magnitude can change upon imposition of an altered potential, the protein will be responsive to the membrane potential. Figure 6.6 illustrates this fact. Imposition of an increased transmembrane electrical potential induces the protein to assume a conformation in which the dipole is altered either in magnitude or direction. If this second protein conformation exhibits altered activity relative to the first, transport function will be influenced by the membrane potential. Responsiveness of a transmembrane protein to the membrane potential is probably a general characteristic, and not a highly selective property of these proteins.

Recent experiments have led to the recognition of a second possible mechanism by which the membrane potential can regulate the activities of permease proteins. The formulation of this mechanism resulted from studies of the Enzymes II of the bacterial phosphotransferase system. As noted in the previous section, the activities of these enzymes are regulated by the membrane potential, and considerable evidence suggests that regulation may involve the reversible oxidation or reduction of thiol groups within the enzymes. This postulate is illustrated in

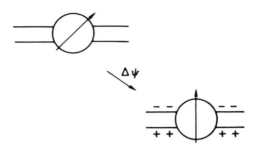

Figure 6.6. Theoretical effect of an imposed membrane potential on the dipole moment of a transmembrane protein. If the electrical dipole of the protein is nonperpendicular to the plane of the membrane, imposition of an electrical potential across the membrane will tend to convert the protein to a conformation with its dipole more perpendicular to the plane of the membrane. If these two conformations differ in biological transport activity, the permease or channel protein will be regulated by the potential. In the cases of the Na^+ and K^+ channels of the nerve membrane, the channels open in response to a membrane potential that is more positive inside and close in response to one that is more negative.

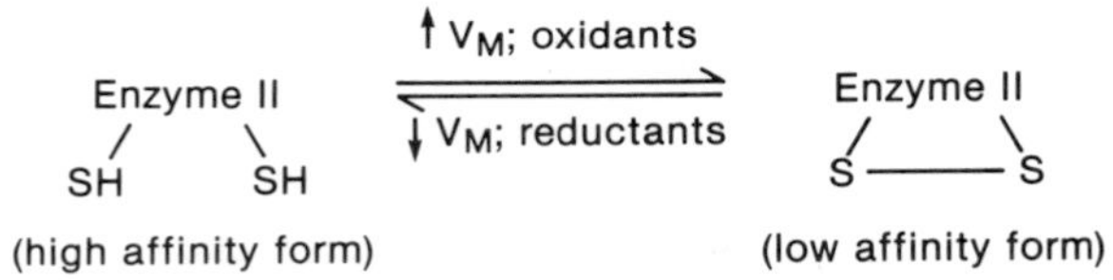

$$
\begin{array}{ccc}
\text{Enzyme II} & \overset{\uparrow V_M;\ \text{oxidants}}{\underset{\downarrow V_M;\ \text{reductants}}{\rightleftharpoons}} & \text{Enzyme II} \\
\diagup \quad \diagdown & & \diagup \quad \diagdown \\
\text{SH} \qquad \text{SH} & & \text{S} \text{———} \text{S} \\
\text{(high affinity form)} & & \text{(low affinity form)}
\end{array}
$$

Figure 6.7. Schematic depiction of the reversible oxidation of the Enzymes II of the bacterial phosphotransferase system. The process is thought to involve the interconversion of a dithiol (reduced) form of the enzyme and a disulfide (oxidized) form of the enzyme. Oxidants such as ferricyanide or oxidized glutathione convert the enzyme to the disulfide form while disulfide reductants such as dithiothreitol or reduced glutathione reverse enzyme oxidation, generating the dithiol form of the enzyme. An increase in the magnitude of the membrane potential (V_M) (negative inside) promotes enzyme oxidation. Sugar binding is thought to promote enzyme reduction because the reduced form of the enzyme possesses high affinity for the substrate. This model provides a possible molecular explanation for the regulation of permease function by the membrane potential.

Figure 6.7. Thiol groups within the enzyme may undergo oxidation with the formation of intramolecular or intermolecular disulfide bridges. This oxidized form of the enzyme apparently possesses low affinity for the sugar substrate relative to the reduced form. An increase in the magnitude of the membrane potential shifts the equilibrium towards the oxidized form. In this way the enzyme may be responsive to the membrane potential and thus responsive to the energy state of the cell. When excess energy is available to the cell and the magnitude of the membrane potential is large, sugars are not readily transported by the PTS. When there is an energy deficiency so that the membrane potential is low, the equilibrium is shifted towards the reduced, maximally active form so that more carbohydrate is transported into the cell. Such a mechanism allows the cell to regulate the quantity of an energy source taken up from the medium in accordance with its needs.

Other permeases may be regulated by an analogous mechanism. For example, the lactose permease may also be responsive to the membrane potential by a mechanism involving sulfhydryl groups. Some permeases may respond with a change in the maximal velocity of transport rather than by altering the affinity of the system for the substrate. Still other transport systems may exist alternatively in an active and an inactive form, one of which is reduced, the other which is oxidized. It seems clear that many variations on this mechanistic theme may be envisaged to accommodate the physiological needs of the cell.

Evolution of Interrelated Metabolic Enzymes, Transport Proteins, Membrane Receptors, and Regulatory Proteins

The preceding discussion revealed that catabolic enzymes, permeases, chemoreceptors, and certain cellular regulatory proteins are related. First, they may share a common general function such as nutrient or energy acquisition, and

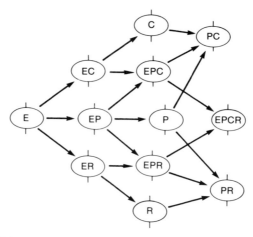

Figure 6.8. Possible pathways for the evolutionary appearance of permeases (P), chemoreceptors (C), regulatory proteins (R), and multifunctional proteins possessing more than one of these functions. All proteins are assumed to be derived originally from metabolic enzymes (E) present in evolving primordial cells. It is proposed that an evolving protein may gain a new function before its original function is lost. Such an event gives rise to multifunctional proteins. Examples of most of the multifunctional proteins depicted are known in present-day organisms. The proteins shown may have evolved as soluble proteins, membrane-associated proteins, or integral constituents of the membrane. Mechanisms allowing evolution of genes coding for proteins of divergent functions may have included (1) nucleotide substitution, (2) gene duplication followed by divergence, and (3) gene fusion and splicing events.

second, they show overlap in their biochemical machinery. For example, the maltose-binding protein (the E protein) functions both in maltose chemoreception and transport, while the cytoplasmic constituent of the maltose permease (the K protein) functions in transport and regulation. The Enzymes II of the PTS not only function in these capacities but also retain enzyme functions, modifying their sugar substrates chemically. These observations are consistent with the hypothesis that metabolic enzymes, permeases, receptors, and regulatory proteins that possess only a single function may have evolved from common ancestral proteins.

Our consideration of transport processes has revealed that several mechanisms coupling energy to transport exist in living cells and that there is more than one mode of solute translocation. These mechanisms probably involve simple and stereospecific protein pores as well as catalytic carriers, which may or may not catalyze chemical reactions involving the transported species. The scheme summarized in Figure 6.8 provides suggestions as to the pathways by which these interrelated proteins may have evolved.

Selected References

Dills, S.S., A. Apperson, M.R. Schmidt, and M.H. Saier, Jr. Carbohydrate transport in bacteria, *Microbiol. Rev. 44:*385 (1980).

Ghosh, B.K. *Organization of Prokaryotic Cell Membranes,* Volume I, CRC Press, Inc., Boca Raton, Florida, 1981.

Heller, K.B., E.C.C. Lin, and T. H. Wilson. Substrate specificity and transport properties of the glycerol facilitator of *Escherichia coli, J. Bact. 144:*274 (1980).

Jacobson, G.R. and M.H. Saier, Jr. "Biological Membranes: Transport" in *Biochemistry* (G. Zubay, ed.), Addison-Wesley Publishing Co., Inc., Reading, Massachusetts, 1983.

Jacobson, G.R. and M.H. Saier, Jr. "Neurotransmission" in *Biochemistry* (G. Zubay, ed.), Addison-Wesley Publishing Co., Inc., Reading, Massachusetts, 1983.

Jacobson, G.R., C.A. Lee, and M.H. Saier, Jr. Purification of the mannitol-specific Enzyme II of the *Escherichia coli* phosphoenolpyruvate:sugar phosphotransferase system, *J. Biol. Chem. 254:*249 (1979).

Kaczorowski, G.J. and H.R. Kaback. Mechanism of lactose translocation in membrane vesicles from *Escherichia coli.* 1. Effect of pH on efflux, exchange, and counterflow. *Biochemistry 18:*3691 (1979).

Kane, J.F., ed. *Multifunctional Proteins: Catalytic/Structural and Regulatory,* CRC Press, Boca Raton, Florida, 1982.

Martonosi, A.N., ed. *Membranes and Transport: A Critical Review,* Vol. I and II, Plenum Publishing Corp., New York, 1982.

Masson, A., ed. *The Maltose System as a Tool in Molecular Biology,* Annals of Microbiology 133, Paris, 1982.

Newman, M.J., D.L. Foster, T. H. Wilson, and H.R. Kaback. Purification and reconstitution of functional lactose carrier from *Escherichia coli, J. Biol. Chem. 256:*11804 (1981).

Oxender, D., A. Blume, I. Diamond, and C.F. Fox, eds. *Membrane Transport and Neuroreceptors,* Progress in Clinical and Biological Research, Vol. 63, Proceedings of the ICN–UCLA Symposium, Alan R. Liss, Inc., New York, 1980.

Robillard, G.T. The enzymology of the bacterial phosphoenolpyruvate-dependent sugar transport system. *Mol. Cell. Biochem. 46:*3–24 1982.

Rosen, B. P., ed. *Bacterial Transport,* Vol. 4, Marcel Dekker, Inc., New York, 1978.

Saier, M.H., Jr. Catalytic activities associated with the enzymes II of the bacterial phosphotransferase system. *J. Supramolec. Struc. 14:*281 (1980).

Saier, M.H. Jr. *Mechanisms and Regulation of Carbohydrate Transport in Bacteria,* Academic Press, in press, 1984.

Saier, M.H., Jr. and C.D. Stiles. *Molecular Dynamics in Biological Membranes,* Springer-Verlag, New York, 1975.

CHAPTER 7

Cell Surface Receptors, Second Messengers, and the Control of Cellular Activities

> In the center there is a narrow path
> that leads to the top of the mountain.
> That path suggests a harmony in life.
> It is the path of perfection.
>
> *Juan Mascaro*

Cells that possess the capacity to respond to external stimuli are socially and environmentally adaptable. If they adapt appropriately, this capacity has enormous survival value. As a result, simple bacteria and unicellular eukaryotes respond to beneficial or deleterious chemicals by swimming towards or away from them, respectively (Chapter 8), while most sexual organisms respond to chemical cues emitted by potential mates. These chemicals signal mating responses, thereby promoting genetic exchange and species continuity (Chapter 9). The mechanisms that provide for adaptive responses probably trigger many developmental processes, from sporulation in bacteria to the most complex sequences of differentiation in higher mammals. In this and the subsequent three chapters, we therefore focus on the mechanisms and control of intercellular signaling, intracellular transmission, and response of the target cellular structure to the transmitted signal. An understanding of the evolution and mechanisms of such communication systems depends on familiarity with the mechanisms of solute recognition and transport discussed in the previous chapter.

Mechanisms of Inter- and Intracellular Signal Transmission

Developing multicellular organisms are highly complex systems in which individual cells communicate with each other by transmitting and receiving signals to and from other elements within the body. They must also be capable of responding to stimuli from external sources. Three types of mechanisms mediate intercellular communication. First, electrical impulses can pass from one cell to another, provided that the two cells exhibit some degree of cytoplasmic continuity. Second, small chemicals such as neurotransmitters and hormones may be released into the extracellular milieu by one cell type and received by another even though the two cells are distant from each other. And third, macromolecular substances may be secreted onto the cell surfaces or into the intercellular matrix where they interact with receptors on the surfaces of other cells, thereby eliciting appropriate responses.

All three mechanisms of communication involve interactions at cellular surfaces. For example, electrical stimuli, passed from one cell to another, may depolarize the receptive cell, thereby causing ion channels within the plasma membrane to open. Neurotransmitters and small hormones relay their messages chemically by binding to cell surface receptors that affect the conformations of membrane enzymes and transport proteins. And macromolecular agents of communication usually interact with complementary macromolecules on the surfaces of the target cells to effect changes in the cellular architecture or metabolism.

In many cases, the reception of electrical, chemical or macromolecular signals at the surface of the target cell is transmitted to the nucleus and other cellular organelles, or to intracellular enzymes. Again, three known mechanisms for the intracellular transmission of information have been elucidated: electrical, chemical, and macromolecular. First, the opening of ion channels both influences the cellular electrical potential and changes the ionic composition of the cytoplasm. Both the membrane potential and local changes in the concentration of a particular ion may be transmitted to other parts of the cell. In muscle cells, for example, membrane depolarization is transmitted throughout the extensive cytoplasmic T system, and this electrical stimulation causes Ca^{2+} channels in the sarcoplasmic reticulum to open and release Ca^{2+}. The increased Ca^{2+} concentration in the cytoplasm brings about the contraction of actomyosin complexes. Consequently, an ion such as calcium can function as a *second messenger,* transmitting information from a primary messenger at the cell surface by a relay mechanism to intracytoplasmic elements. Examples of ionic transmission mechanisms in simple systems are considered in Chapters 2 (*Fucus*), 8 (*Paramecium* and *Dictyostelium*), and 9 (*Bombyx*).

The second transmission mechanism, which is well established in most eukaryotes, involves synthesis at the cell surface of a small regulatory nucleotide. Hormone reception may enhance (or diminish) the activity of a membrane-bound cyclic nucleotide biosynthetic enzyme via an intramembrane transducer protein. The change in the cyclic nucleotide biosynthetic rate alters the cytoplasmic concentration of this compound that is sensed by other catalytic elements in the cell. The cyclic nucleotides therefore represent a second type of second messenger. Examples of cyclic nucleotide mediated transmission mechanisms in microorganisms are discussed in Chapters 2 (*Caulobacter*), 8 (*Dictyostelium*), and 9 (*Saccharomyces*).

The third general transmission mechanism involves direct protein–protein interaction. In this case, the cell surface receptor either controls the cytoplasmic concentration of a soluble protein, or modifies it so that its regulatory characteristics change. While second messenger proteins are presently less widely recognized as mediators of intracellular communication, their widespread roles in the regulation of cellular metabolic activities and gene function are likely to become universally appreciated as further investigations are conducted. Signal transmission from receptor to flagellum during bacterial chemotaxis (Chapter 8) may occur by such a mechanism.

In this chapter we focus on some of the best characterized examples of macromolecular transmission. We shall see that such mechanisms can control the activities of permeases and enzymes and determine the frequencies of transcriptional initiation by RNA polymerase. Interrelationships between macromolecular and micromolecular transmission mechanisms will become apparent.

Control of Carbohydrate Catabolic Enzyme Synthesis by Intracellular Inducer and Cyclic AMP

Even before the turn of the century, it was known that when microorganisms were grown in the presence of two different carbon sources, one of these was frequently utilized preferentially. For example, when *E. coli* cells are grown in the presence of both glucose and lactose, the glucose is metabolized first, and lactose catabolism begins only after the glucose has been removed from the medium. Glucose functions in this capacity by reducing the concentrations of the enzymes involved in lactose utilization. The phenomenon is known as the *glucose effect* or *catabolite repression* and is limited to certain carbohydrate permeases and degradative enzymes.

Extensive studies have shown that synthesis of a carbohydrate metabolic enzyme system, such as that involved in lactose breakdown in *E. coli*, is dependent on the presence of two small cytoplasmic molecules, inducer and cyclic AMP. A model depicting the interactions of regulatory proteins with the controller region of the *lac* (lactose) operon is shown in Figure 7.1. These three proteins are: RNA polymerase, which transcribes the DNA sequence of the operon into messenger RNA, and two regulatory proteins that determine the frequency with which messenger RNA synthesis is initiated, the *lac* repressor and the cyclic AMP receptor protein (CR protein). The latter two proteins can each exist in either of two conformations, only one of which can bind to the DNA. The relative proportions of these two conformations are controlled by two small cytoplasmic ligands, inducer and cyclic AMP, respectively. The free form of the *lac* repressor binds to the *lac* operator region of the operon to prevent transcription, but inducer molecules bind to the repressor protein, converting it to a form that possesses low affinity for the DNA. Consequently, an elevated intracellular concentration of inducer causes the *lac* repressor protein to dissociate from the *lac* operon. By contrast, the free form of the cyclic AMP receptor protein possesses low affinity for the controller region of the operon, but the binding of cyclic AMP to the protein converts it to a conformation with high affinity for the *lac* promoter. Thus, an enhanced intracellular concentration of cyclic AMP causes the CR protein to associate with the DNA. But since the binding of the cyclic AMP–CR protein complex to the promoter promotes transcription (positive control),

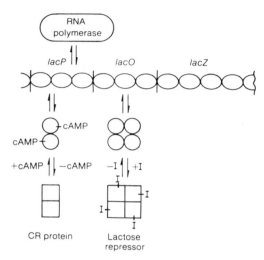

Figure 7.1. Proposed mechanism of transcriptional regulation of the lactose operon in *E. coli*. Reproduced from M.H. Saier, Jr. and C.D. Stiles, *Molecular Dynamics in Biological Membranes*, Springer-Verlag, 1975, with permission.

while the binding of the *lac* repressor to the operator inhibits transcription (negative control), an elevation in the concentration of either cyclic AMP or inducer should enhance the rate of *lac*-specific messenger RNA synthesis. Expression of the operon is clearly under dual control by the two small cytoplasmic molecules, inducer and cyclic AMP.

If carbohydrate catabolic enzyme systems are generally under dual control by inducer and cyclic AMP (and extensive evidence supporting this notion is available), then the bacterium might benefit from a mechanism that results in the coordinate regulation of the intracellular levels of these two molecules. In fact, the transport systems responsible for the uptake of several inducers and the cyclic AMP synthetic enzyme, adenylate cyclase, are subject to coordinate regulation. Thus, when glucose or mannitol is added to a culture of growing *E. coli,* inducer uptake and cyclic AMP synthesis are immediately inhibited. As a consequence of extensive genetic, physiological, and biochemical studies, a specific regulatory mechanism involving the proteins of the bacterial phosphotransferase system (Chapter 6) has been proposed.

Coordinate Regulation of Adenylate Cyclase and Inducer Uptake by the Phosphotransferase System

It is worth reemphasizing that the phosphotransferase system consists, in essence, of a phosphate transfer chain; the phosphoryl moiety of phosphoenolpyruvate is first transferred to Enzyme I with the formation of a high-energy phosphoryl protein in which the phosphate group is covalently linked to a histidyl residue in the protein. Second, the phosphoryl moiety is transferred from phosphoenzyme I to the heat-stable protein, HPr, with the formation of a second high-energy phosphoprotein. As in the case of Enzyme I, a histidyl residue in HPr is phosphorylated. The third phosphoryl transfer reaction in this sequential progression of events involves transfer of the phosphoryl moiety from the general, non-sugar-specific protein, HPr, either to a sugar-specific Enzyme III such as the glucose Enzyme III (Enzyme III^{glc}) or to an Enzyme II such as the mannitol Enzyme II (Enzyme II^{mtl}). The transport of glucose across the membrane involves the transfer of the phosphoryl moiety of Enzyme III^{glc} to Enzyme II^{glc}, and then to the sugar. This scheme is shown in Figure 6.2.

How the proteins of the phosphotransferase system are thought to function in regulating adenylate cyclase and the carbohydrate permeases is shown in Figure 7.2. In a cell that is not in contact with a sugar substrate of the PTS, phosphoenolpyruvate will phosphorylate all of the PTS proteins: Enzyme I, HPr, and the sugar-specific Enzymes III and II. If a sugar such as glucose or mannitol is added to the medium, the sugar will be transported and phosphorylated, thereby draining phosphate off of the high energy phosphoproteins.

Regulation of permease activity is thought to be effected by an allosteric regulatory protein, termed RPr (Figure 7.2). The binding of RPr to the cytoplasmic surface of the active permease converts it to a conformation that has low transport

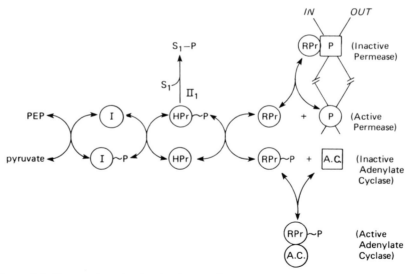

Figure 7.2. Proposed mechanism for the coordinate regulation of carbohydrate permeases and adenylate cyclase in *E. coli*. AC, adenylate cyclase; RPr, regulatory protein; S, sugar substrate of PTS. From M.H. Saier, Jr. in *The Bacteria*, Vol. VII (I.C. Gunsalus, J.R. Sokatch, and L.N. Ornston, eds.), Academic Press, 1979, reprinted with permission.

activity. Only the free form of RPr, not the phosphorylated form, binds to the permease.

Adenylate cyclase is also thought to be regulated by RPr but in a somewhat different fashion. The free form of RPr is not believed to bind to the adenylate cyclase complex. Instead, the phosphorylated form may bind and activate an enzyme that otherwise possesses low activity. Thus, while the permeases are subject to negative control by free RPr, adenylate cyclase may be subject to positive control by phospho-RPr.

As noted above, addition of a sugar substrate of the phosphotransferase system to the medium of a growing bacterial culture results in the immediate group translocation of the sugar into the cell. The phosphoryl groups of the phospho-proteins are transferred to the incoming sugar. Because all of the phosphotrans-ferase phosphoproteins are of high energy and in equilibrium with each other, removal of the phosphoryl group from one protein (e.g., phospho-HPr) should remove phosphate from the others (e.g., RPr). Addition of sugar therefore con-verts phospho-RPr to free RPr, and this interconversion results in the deactivation of adenylate cyclase as the concentration of phospho-RPr decreases, and the inhibition of the permeases as the concentration of free RPr increases.

Genetic and biochemical experiments have shown that RPr is the glucose Enzyme III, a protein that is synthesized in invariant amounts. Mutants lacking Enzyme III^{glc} exhibit very low adenylate cyclase activity, and the permeases are not subject to inhibition by sugars. Further, when Enzyme III^{glc} is placed into membrane vesicles derived from *E. coli* cells, lactose uptake is inhibited. Phos-phorylation of the Enzyme III restores transport activity fully. This functional regulation has been demonstrated in a reconstituted protein–phospholipid vesicle system consisting of only phospholipids, the lactose permease protein, and the proteins of the PTS.

Another experimental approach has been to demonstrate the direct binding of Enzyme III^{glc} to the lactose permease protein *in vitro*. While binding of free Enzyme III^{glc} to the permease could be easily demonstrated, phospho-Enzyme III^{glc} did not bind. It was further found that Enzyme III^{glc} bound to the permease only if lactose, or another sugar substrate of the permease was present. In other words, lactose and Enzyme III^{glc} binding exhibit cooperativity. The binding of one promotes binding of the other. These interrelationships are illustrated in Figure 7.3.

Cooperative binding of the sugar substrate and the allosteric regulatory protein to the permease is undoubtedly of physiological significance. Because Enzyme III^{glc} must function in glucose uptake as well as regulate several permeases and adenylate cyclase, it would be wasteful to bind Enzyme III^{glc} to a permease that is nonfunctional because of a lack of substrate. As a consequence of cooperativity the cytoplasmic Enzyme III^{glc} is used only where and when it is needed.

It is valuable to consider this regulatory mechanism in terms of the receptor–second messenger concepts expounded at the beginning of the chapter. The glucose or mannitol receptor is the Enzyme II^{glc} or Enzyme II^{mtl}, respectively,

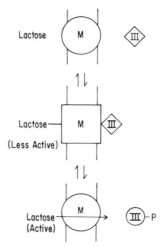

Figure 7.3. Proposed mechanism of the regulatory interaction between Enzyme III[glc] and the lactose permease. M and III indicate lactose permease (M protein) and Enzyme III[glc], respectively. Lactose and Enzyme III[glc] are shown to bind to the lactose permease on opposite sides (outer surface for lactose and inner surface for Enzyme III[glc]) of the plasma membrane (the area between the two vertical lines). The cooperative conformational change induced by either sugar or Enzyme III[glc] binding can therefore be thought of as a transmembrane signaling device. From T. Osumi and M.H. Saier, Jr., *Proc. Natl. Acad. Sci. USA* 79:1457–1461 (1982), with permission.

which functions together with the sugar to control the phosphorylation state of Enzyme III[glc]. Enzyme III[glc] is the second messenger, which transmits information from the phosphotransferase system receptors, through the cytoplasm to other cellular constituents, namely, the target permeases and adenylate cyclase. That is, the effector or sugar-sensing receptor is an Enzyme II of the phosphotransferase system. The second messenger is a cytoplasmic protein of invariant concentration that can exist in alternative conformations determined by its phosphorylation state; the targets of second messenger control are adenylate cyclase and various sugar permeases. This regulatory system probably represents one of the best characterized of all intracellular macromolecular receptor-transmission signaling devices yet studied.

Induction of Protein Synthesis by Extracellular Inducers

In the previous discussion, a protein–protein receptor–transmission system was found to regulate the activities of the enzyme and permeases which respectively control the cytoplasmic concentrations of cyclic AMP and inducer. These small molecules in turn regulate the rates of transcription of carbohydrate catabolic operons such as the lactose operon in *E. coli*. Two other possible examples of protein mediated receptor-transmission systems controlling bacterial operon expression are discussed in this section. Syntheses of a number of proteins that function as components of certain permeases are induced only when the inducers are bound to receptors on the external surface of the cell membrane. Among the proteins whose syntheses are controlled by exogenous inducers are some of the proteins of the phosphotransferase system and a permease complex responsible for the uptake of hexose phosphates. Extracellular induction mechanisms apparently evolved for those permeases that transport substances representing normal metabolic intermediates found in cells. If these intracellular compounds were to serve as inducers, synthesis of the permeases would occur even when the substrates were absent from the external medium. The cell would waste energy synthesizing unnecessary proteins, and the permeases might "leak" nutrients to the external environment. These considerations necessitated the evolution of novel transcriptional regulatory mechanisms such as those described below.

Two separate mechanisms have been proposed for the regulation of permease protein synthesis by exogenous inducers. These two mechanisms represent processes that may be operative for induction of the *pts* operon (coding for the energy coupling proteins of the PTS, Enzyme I, and HPr) and for induction of the *uhp* operon, coding for the protein(s) of the hexose phosphate permease. Studies of the former system led to the following observations:

(1) Induction was dependent on an extracellular sugar substrate of the PTS, as well as the Enzyme II specific for the inducing sugar. Genetic loss of the Enzyme II resulted in a noninducible phenotype with respect to that particular sugar substrate.

(2) Induced synthesis was dependent on cyclic AMP and a functional cyclic AMP receptor protein. In the absence of either agent, the rate of protein synthesis was basal or subbasal.

(3) Several extracellular substrates of the phosphotransferase system induced expression of the *pts* operon, and the best inducers were the most rapidly transported substrates of the system. Poor substrates were poor inducers.

(4) Loss of either Enzyme I or HPr activity as a result of mutations in the structural genes for these proteins gave rise to high-level constitutive synthesis of the other protein. These results have led to the postulation of a mechanism involving phosphorylation of a regulatory protein, which can interact with the DNA to regulate transcription (Figure 7.4). In this proposal, phosphorylated RPr is a transcriptional repressor (negative control [Figure 7.4, left]) and/or free RPr is an activator (positive control [Figure 7.4, right]). The RPr in Figure 7.4 is not necessarily the same protein as the RPr in Figure 7.2.

The second well-characterized, exogenously induced permease system is the hexose phosphate permease, encoded by the *uhp* operon. Mutant analyses, as well as limited biochemical studies have led to the suggestion that the model depicted in Figure 7.5 explains induction of this operon. Here, the inducer, glucose-6-phosphate, is presumed to bind to the external surface of the cytoplasmic membrane, by virtue of the presence of a transmembrane receptor (MR).

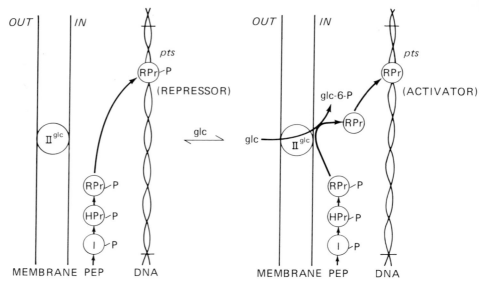

Figure 7.4. Proposed mechanism for the regulation of transcription of the *pts* operon in *E. coli*. The model suggests that a regulatory protein (here designated RPr), which can be phosphorylated at the expense of phospho-HPr, can interact directly with the operator region of the *pts* operon. Phospho-RPr may be a repressor protein, and/or free RPr may be an activator. The addition of a substrate of the PTS to the cell suspension would be expected to drain phosphate off of the phosphorylated energy-coupling proteins of the PTS with concomitant phosphorylation of sugar. Hence, phospho-RPr would be dephosphorylated to RPr, and transcription of the *pts* operon would be induced. Abbreviations: glc, glucose; II^{glc}, Enzyme II specific for glucose; PEP, phosphoenolpyruvate; glc-6-P, glucose 6-phosphate; I, Enzyme I; DNA, deoxyribonucleic acid. From S.S. Dills, A. Apperson, M.R. Schmidt, and M.H. Saier, Jr., *Microbiol. Rev. 44*:385–418 (1980), copyright American Society for Microbilogy, reprinted with permission.

When glucose-6-phosphate (G6P) is bound to the receptor, a conformation of the receptor is induced that possesses high affinity for the repressor of the *uhp* operon. This cytoplasmic protein is largely removed from the cytoplasm, and therefore dissociates from the operator site of the *uhp* operon. Dissociation of the repressor from the DNA allows transcription of the *uhp* operon by RNA polymerase, and hence synthesis of the hexose phosphate permease is induced.

These two proposed mechanisms of transcriptional regulation represent *catalytic* versus *structural* mechanisms for regulating transcription by protein second messengers. The *pts* operon is presumed to be regulated by a catalytic mechanism because transcription depends on the reversible, enzymatic phosphorylation of the repressor, RPr. By contrast, the *uhp* operon is regulated by a structural mechanism, because the transcriptional repressor is not modified by a covalent chemical reaction. Instead, a change in the cellular location of the protein is proposed to be responsible for the activation of gene expression.

These two transcriptional regulatory processes represent transmission mech-

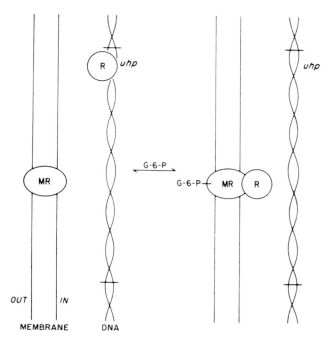

Figure 7.5. Possible mechanism for the regulation of gene expression by exogenous inducer. The model suggests an explanation for transcriptional regulation of the protein or proteins of the hexose phosphate transport system in *E. coli*. Two regulatory proteins are proposed. One is a repressor protein (R), which possesses affinity for the controller region of the operon coding for the transport protein (or proteins) of the system. It also binds to a transmembrane regulatory protein (MR), which possesses a glucose 6-phosphate (G-6-P) binding site on the external surface of the cytoplasmic membrane. The binding of glucose 6-phosphate to the MR protein determines its affinity for the R protein. There are no data available that distinguish this model from one in which a cytoplasmic activator protein (instead of a repressor protein) controls expression of the operon (see text). DNA, Deoxyribonucleic acid. From M.H. Saier, Jr. in *The Bacteria,* Vol. VII (I.C. Gunsalus, J.R. Sokatch, and L.N. Ornston, eds.), Academic Press, 1979, reprinted with permission.

anisms by which a signal at the cell surface is conveyed to the nucleoid region within the cytoplasm. No small cytoplasmic molecules are thought to be involved. These processes, demonstrated in simple bacteria, may therefore serve as models for certain cell surface receptor-mediated processes in higher organisms. The control of transcriptional activities by extracellular hormones, macromolecules and cell–cell contacts during differentiation of multicellular organisms might reasonably be effected by similar mechanisms and may be evolutionarily related to extracellular induction in microorganisms. More information will be required to establish the molecular mechanisms involved. But the establishment of the transmission mechanisms in simple microbes will undoubtedly lead to a better understanding of the corresponding processes in multicellular organisms.

Selected References

Braun, V. and K. Hantke. "Bacterial Cell Surface Receptors" in *Organization of Pro-karyotic Cell Membranes,* Vol. II (B.K. Ghosh, ed.), CRC Press, Boca Raton, Florida, 1981.

Dills, S.S., A. Apperson, M.R. Schmidt, and M.H. Saier, Jr. Carbohydrate transport in bacteria, *Microbiol. Rev., 44:*385 (1980).

Dills, S.S., M.R. Schmidt, and M.H. Saier, Jr. Regulation of lactose transport by the phosphoenolpyruvate-sugar phosphotransferase system in membrane vesicles of *Escherichia coli, J. Cell. Biochem. 18:*239 (1982).

Kane, J.F., ed. *Multifunctional Proteins: Catalytic/Structural and Regulatory,* CRC Press, Boca Raton, Florida, (1983).

Leonard, J.E., C.A. Lee, A.J. Apperson, S.S. Dills, and M.H. Saier, Jr. "The Role of Membranes in the Transport of Small Molecules" in *Organization of Prokaryotic Cell Membranes,* Vol. I (B.K. Ghosh, ed.), CRC Press, Boca Raton, Florida, 1981.

Martonosi, A.N., ed. *Membranes and Transport: A Critical Review,* Vol. I and II, Plenum Publishing Corp., New York, 1982.

Osumi, T. and M.H. Saier, Jr. Regulation of lactose permease activity by the phos-phoenolpyruvate-sugar phosphotransferase system: evidence for direct binding of the glucose-specific enzyme III to the lactose permease, *Proc. Natl. Acad. Sci. USA 79:*1457 (1982).

Saier, M.H., Jr. "The Role of the Cell Surface in Regulating the Internal Environment" in *The Bacteria,* Vol. VII, Mechanisms of Adaptation (I.C. Gunsalus, J.R. Sokatch, and L.N. Ornston, eds.), Academic Press, New York, 1979.

Saier, M.H., Jr. Catalytic activities associated with the enzymes II of the bacterial phosphotransferase system, *J. Supramolec. Struc. 14:*281 (1980).

Saier, M.H., Jr. *Mechanisms and Regulation of Carbohydrate Transport in Bacteria,* Academic Press, in press, 1984.

Saier, M.H., Jr., and E.G. Moczydlowski. "The Regulation of Carbohydrate Transport in *Escherichia coli* and *Salmonella typhimurium*" Chap. 3 in *Bacterial Transport,* Vol. 4 (B.P. Rosen, ed.), Marcel Dekker, Inc., New York, 1978.

Saier, M.H., Jr. and C.D. Stiles. *Molecular Dynamics in Biological Membranes,* Springer-Verlag, New York, 1975.

CHAPTER 8

Mechanisms of Chemoreception, Electrical Signal Transduction, and Biological Response

> Only the day dawns
> to which we are awake
>
> *Henry David Thoreau*

Living organisms receive and respond to a variety of chemicals and energy stimuli, and these interactions influence differentiation and regulate sexual activities. Well-characterized examples include responses of coelenterates to peptide morphogens (Chapter 5) and of haploid yeast cells to sexual pheromones (Chapter 9). In the developing embryo as in developing microbial systems, cell migration and elongation may be governed by chemotactic processes (Chapters 1, 2, and 4). Bioelectric stimuli also play an important role in developmental processes. Examples include the development of cell polarity and tissue regeneration, as discussed in Chapter 2, and fertilization of sea urchin eggs (Chapter 10). Thus, it is crucial to an understanding of differentiation that the mechanisms of chemoreception and electrical signal transmission be elucidated. In this chapter we discuss chemotactic and mechanotactic processes in Gram-negative bacteria and selected eukaryotic microorganisms.

Bacterial Chemotaxis: A Model of Sensation, Regulation, and Adaptation

Chemotaxis in bacteria involves elements of sensory reception and regulatory biochemistry that are integrated to produce a behavioral response in a single cell. In many respects, this system appears to resemble hormonal regulation in higher organisms, in which membrane-associated receptors and second messengers are used for detection of chemical signals resulting in appropriate cellular responses (see Chapter 7). Furthermore, bacteria can "adapt" to sensory stimuli, allowing the organism to perceive relatively small changes in environmental factors over a large range of absolute levels of these stimuli. The advantages of the bacterial system for use in studying such phenomena, however, lie in the ease of obtaining mutants in the detection and signal processing components, and in one's ability to obtain large amounts of a single type of cell from which to isolate and study the cellular components involved. For these reasons, bacterial chemotaxis has become a popular system in which to study the biochemical bases of sensation, regulation and adaptation in the behavioral responses of cells.

Chemotaxis in bacteria was first described about a century ago by T.W. Engelmann and W. Pfeffer. They observed that motile bacteria that swim by means of one or more flagella can respond to gradients of chemical substances present in their environment by swimming toward or, in some cases, away from regions containing higher concentrations of the compound. Chemoattractants include such nutrients as amino acids, sugars, and oxygen while repellents include substances detrimental to the cell such as organic acids and alcohols, and extremes of pH.

The chemotactic process has been shown to require cellular components that act as chemoreceptors, information processors, and response regulators, which ultimately convert a chemical stimulus into a modulation in the motion of the bacterial flagella. Consequently, before we can understand the chemotactic process, we must have some idea of the mechanism of bacterial motility.

In electron micrographs of *E. coli*, flagella appear as filaments approximately 20 μm long and 20 nm in diameter (Figure 8.1). These filaments are helical structures consisting of a single protein, flagellin. Flagellar filaments isolated by mechanical shearing techniques appear uniform and consist almost exclusively of the protein flagellin. However, when Gram-negative bacterial cells are disrupted by treatment with detergents, the isolated flagella exhibit a more complex structure at one end consisting of a 0.9 μm long hook and a *basal body*, a thin rod to which are attached four rings. The rod and associated rings are embedded in the cell wall of undisrupted *E. coli*; the two outer rings are believed to interact with the lipopolysaccharide and peptidoglycan layers of the cell envelope, while the inner two rings are associated with the cytoplasmic membrane.

The roles of these complex basal structures of bacterial flagella can be inferred from the mechanism by which the flagella propel the cell. Prokaryotic flagella rotate at their base. Rotation occurs at 10–20 revolutions per second and may be clockwise or counterclockwise depending on stimuli in the surrounding medium. The suggested roles for the rings are to act as *stators* and *rotors* for the flagellar rotary motor. Rotation of these structures with respect to one another could then generate the torque necessary to spin the flagellar filament. It is believed that the proton electrochemical gradient provides the energy for flagellar rotation.

The motion of *E. coli* cells in a medium of uniform composition is characterized by a series of smooth runs in a single direction, lasting about one second, punctuated by episodes of tumbling, which last about 0.1–0.2 sec (Figure 8.2A). After a tumble, the organism takes off again on a smooth run in a new, randomly chosen direction. In the presence of a gradient of a chemoattractant (or chemorepellent), however, the motions of bacteria become biased. The organisms tumble less frequently when swimming toward a region of higher attractant concentration (or away from a repellent), and tumble more frequently if they should happen to choose a direction that leads them away from the compound

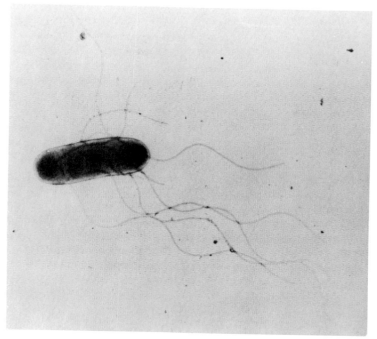

Figure 8.1. Electron micrograph of a *Salmonella typhimurium* cell showing helical flagella inserted at various places along the cell surface. Such a mode of flagellar insertion is termed peritrichous. Other bacteria exhibit polar flagellation, with flagellar insertion sites only at one end of the cell. (From D.E. Koshland, Jr., *Physiol. Rev. 59*:811–862 (1979), reprinted with permission).

(or toward the compound in the case of a repellent) (Figure 8.2B). This behavior explains the ability of these organisms to swim toward regions that are more beneficial (or less hostile) to their survival.

We now know that a transient increase in attractant concentration favors counterclockwise flagellar rotation, while clockwise rotation is stimulated by addition of a repellent or a rapid decrease in an attractant concentration. Smooth swimming is the result of prolonged counterclockwise rotation, while a switch in the direction of rotation causes tumbling. Normally, *E. coli* cells that are swimming smoothly have flagellar filaments that are in a left-handed helical configuration and form a compact bundle at one end of the cell as they rotate counterclockwise (Figure 8.3A). A tumble occurs when the direction of rotation is abruptly reversed, and the helical filaments fly apart because the left-handed helices are unable to rotate in concert within the bundle when they are forced to undergo clockwise rotation (Figure 8.3B). Thus, chemoeffectors influence the direction of flagellar rotation which, in turn, determines the swimming behavior of the cell.

As long as bacterial cells experience an increase in attractant concentration,

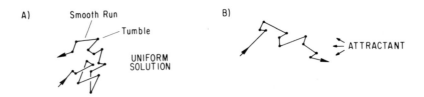

Figure 8.2. Hypothetical tracks of an *E. coli* cell in a uniform medium (A), and a medium containing a gradient of chemoattractant (B). In A, smooth runs of about 1 sec (average) in random directins are punctuated by 0.1–0.2 sec tumbles. The individual cell therefore carries out a "random walk." In B, tumbling is suppressed if the cell is swimming up the concentration gradient of attractant and stimulated if it selects a direction away from the compound. This leads to a biased random walk, the net result of which is to lead the organism toward higher concentrations of the chemoattractant.

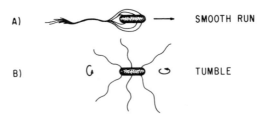

Figure 8.3. Physical explanation of smooth runs and tumbles. During smooth runs, counterclockwise rotation of left-handed helical flagella results in a flagellar bundle at one end of the cell that propels the organism in one direction (A). A brief, abrupt switch to clockwise rotation causes the individual filaments in the bundle to fly apart, and the cell undergoes a tumble (B).

tumbling is suppressed. When they reach a region in which a gradient no longer exists, they adapt or revert to normal swimming behavior. Experiments in which the attractant concentration is constantly increased in a uniform manner in the medium without the production of spatial gradients have shown that the tumbling frequency is continuously suppressed during this period. Therefore, bacteria sense a temporal change in chemoeffector concentration. They are able to "compare" the level of attractant at one point in time to that which they sensed a short time earlier. If the new level is greater, they continue to swim smoothly. This rudimentary "memory," however, is short lived. If further increases in attractant concentration are not encountered, the organism forgets within a few minutes that at an earlier time the levels were lower, and it resumes its random walk.

Bacterial Chemoreceptors

The mechanisms by which chemotactic effectors are sensed and converted into responses by the bacterial cell have been probed by a combination of genetic and biochemical techniques. In most cases, there appears to be a primary chemotactic receptor that interacts with the attractant and transmits the information that an effector is present to a processing system affecting flagellar rotation. For the sugars, maltose, ribose, and galactose, the primary receptors in *E. coli* are periplasmic binding proteins that are also involved in the transport of these carbohydrates into the cell (Chapter 6). In contrast, the receptors for the amino acids aspartate and serine are believed to be 60,000-dalton transmembrane proteins that are products of the *tar* and *tsr* genes, respectively. It should be pointed out that although many of the receptors identified so far are involved in transport of the solutes they recognize, transport is not always a prerequisite for chemotaxis. There are mutants in which transport of a solute is eliminated, but chemotaxis toward it still occurs.

In the case of those compounds recognized by periplasmic binding proteins, the *signal* is transmitted to integral membrane proteins by the binding protein–solute complex. For galactose and ribose, this protein is the product of the *trg* gene in *E. coli,* which also has a molecular weight of 60,000. For maltose, the secondary receptor appears to be the *tar* gene product, which interacts with the maltose-binding protein–maltose complex as well as with free aspartic acid. The flow of information to integral membrane receptors during the initial steps of chemotactic reception is schematically depicted in Figure 8.4. Ultimately, the signal is transmitted to the transmembrane *tar, trg,* or *tsr* gene products either directly by binding of the effector itself, or indirectly by binding of the binding protein–effector complex. Other chemoattractants and repellents have also been shown to feed information to one of these three proteins as shown in Figure 8.4. It is worth noting that all three of these proteins have been found to exhibit similar sizes and amino acid sequences, suggesting that they evolved from a common ancestral protein. This fact suggests that they function by means of a uniform mechanism. Thus, chemotactic information is focused into only a few transmembrane proteins, which then transmit an integrated signal to the flagellar apparatus.

Clues to the functions of the *tar, trg,* and *tsr* proteins in chemotaxis have come from a number of observations. Mutations that eliminate only one of these proteins abolish chemotaxis to only those compounds that feed information to it. In fact, this genetic evidence helped establish the scheme of information flow shown in Figure 8.4. Therefore, these proteins are indispensable links in the excitation phase of the sensory process. A second important observation is that these membrane proteins are methylated in a reaction dependent on S-adenosylmethionine. This reaction, catalyzed by a methyltransferase, involves formation of methyl esters of several glutamic acid residues within each of the *tar, trg,* and *tsr* gene products. In the absence of chemoeffectors, these proteins are in an intermediate state of methylation. Addition of an attractant to *E. coli*

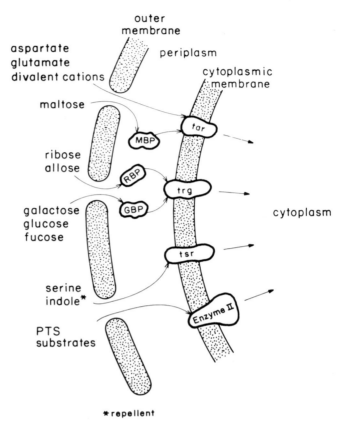

*repellent

Figure 8.4. Information flow through integral membrane proteins in bacterial chemotactic reception for some of the attractants and repellents recognized by *E. coli*. For some compounds, such as aspartate, serine, and the sugar substrates of the phosphotransferase system, the effector interacts directly with cytoplasmic membrane receptors. For others, such as maltose, ribose and galactose, the primary receptor is a periplasmic binding protein, and the binding protein–effector complex binds to the membrane receptor. More than one compound can feed information to each integral membrane protein. Thus, ribose and galactose compete for the same protein (*trg* product) even though they use different primary receptors. An integrated signal is then transmitted from the membrane receptors to the flagellar apparatus (see text).

stimulates methylation of one of these proteins, while repellents cause demethylation, catalyzed by a methylesterase. Furthermore, mutants defective in either the methyltransferase or the methylesterase exhibit aberrant chemotactic responses. These results show that methylation is an important process for regulating chemotactic responses, and these proteins have been called *methyl-accepting chemotaxis proteins* or MCPs.

Genetic techniques have been useful in dissecting the steps necessary for proper processing of the information received by the primary receptors and MCPs. Mutant bacteria that are generally defective in chemotaxis (*che* mutants) presumably lack cellular components central to the conversion of a sensory stimulus into a behavioral response. For example, *cheR* and *cheB* mutants of *E. coli* lack the methyltransferase and methylesterase activities respectively, while the products of other *che* genes are found in the cytoplasm and are presumably involved in transmitting signals from the MCPs to the flagella. Finally, the products of the genes, *cheC* and *cheV*, appear to be flagellar proteins, which receive the chemotactic signals and somehow modulate flagellar rotation in response to them. Based on these observations, an overall scheme of information flow in bacterial chemotaxis can be drawn as shown in Figure 8.5.

Methylation of the MCPs appears to have a role in the adaptation of cells to chemical stimuli. The methylation level is apparently a measure of the absolute concentration of the effector in the medium. Rapid addition of an attractant causes a transient increase in smooth swimming, while a temporary increase in tumbling is the result of a sudden addition of repellent to the medium. The time required for the bacteria to return to normal swimming behavior after these changes correlates with the time course of methylation (attractant addition) and demethylation (repellent addition), respectively. Therefore, *adaptation* to a single stimulus seems to be accomplished by chemical modification of the MCPs to a new, steady state level of methylation.

The observations described above show that MCP methylation cannot be the excitatory signal detected by the flagellar motor, but instead correlates with the adaptation phase. In fact, some mutants in the methylation–demethylation machinery (i.e., *cheR* and *cheB*) can still respond to sudden additions of attractants

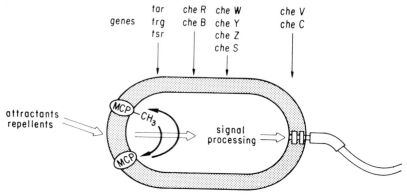

Figure 8.5. Overall information flow in *E. coli* chemotaxis. The cell envelope is depicted as a single layer, and only a single flagellum is shown for simplicity. The levels of participation of various gene products in reception and processing are shown by the arrows above the cell. Signal processing is proposed to affect the level of an unidentified response regulator that may interact with a flagellar protein (possibly the *cheV* and/or *cheC* gene products) to affect the rotational direction of the organelle. (Adapted from Koshland, D.E., Jr., *Ann. Rev. Biochem. 50:*765–782 (1981).)

or repellents, but cannot adapt to them and thus exhibit altered chemotactic behavior. This fact demonstrates that the processes of *excitation* and *adaptation* must be separate, albeit related responses. Both depend on the methyl-accepting chemotaxis proteins, but only adaptation depends on methylation.

Various models have been suggested for how these two processes are interrelated. Most of them have in common a molecule called the *response regulator* (also called the rotation or tumble regulator). Addition of an attractant to a motile culture of *E. coli* is proposed to *transiently* increase the intracellular level of the response regulator. Increased methylation of an MCP then ensues, and once a new methylation level is attained, the concentration of the response regulator returns to its normal value. Repellent addition, or a sudden decrease in attractant concentration, is hypothesized to have the opposite effect. The response regulator would then have a direct role in excitation of the flagellar apparatus. High levels would favor smooth swimming, and lower than normal values would induce a high frequency of tumbling.

The model outlined above explains how multiple chemotactic signals can be integrated into a coherent response by the cell based on the level of a single regulatory molecule. It is important to note, however, that not all chemotactic responses are mediated by the MCPs or by methylation reactions. Chemotaxis toward sugar substrates of the phosphotransferase system (Chapter 6) or oxygen appears to be independent of the MCPs and a methylation process. Some evidence suggest that chemotaxis towards PTS sugars occurs by a mechanism involving the central regulatory protein, RPr, of the PTS (see Chapter 7). In this case, an increase in the concentration of the PTS substrate enhances the rate of sugar uptake and causes the RPr-phosphate/RPr ratio to shift toward free RPr. If the regulatory protein (or its phosphorylated derivative) binds to the basal region of the flagellum to influence its direction of rotation, this event might correspond to excitation. By this means, RPr-dephosphorylation, caused by an increased concentration of extracellular sugar, could induce counterclockwise rotation of the flagellum and smooth swimming. Adaptation might be brought about by some independent, and as yet unidentified process. We expect that multiple mechanisms of excitation and adaptation will come to light in the near future.

The Avoidance Reaction of *Paramecium*

The elongated, ciliated protozoa of the genus *Paramecium* are single-celled, free-living organisms that show a high degree of structural complexity (Figure 8.6). The anterior of the cell has a "mouth" or *oral groove* for ingesting food, while the posterior is characterized by an "anal pore," which is involved in the elimination of cellular wastes. The entire cell surface is covered by approximately 10,000 cilia, the structures responsible for cell locomotion. These are arranged in rows and are anchored to a cytoskeleton beneath the cell surface. All of the cilia and the intervening regions of the cell surface are surrounded by a continuous

Anterior

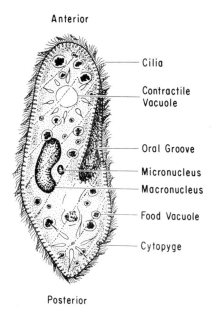

Cilia

Contractile
Vacuole

Oral Groove

Micronucleus

Macronucleus

Food Vacuole

Cytopyge

Posterior

Figure 8.6. A *Paramecium* cell showing various organelles and the oral groove on the anterior half and an anal pore (cytopyge) near the posterior of the cell. The protozoan is covered by rows of cilia, but only those at the margins of the organism are shown in this diagram for clarity. The cell is drawn at approximately 200× its actual size.

plasma membrane; the characteristic shape is conferred upon the cell by the underlying cytoskeleton. By coordinating its beating cilia, *Paramecium* can move forward or backward at a more or less constant rate.

In 1906, H.S. Jennings first described an "avoidance reaction" in *Paramecium* which involves a quick change in the direction of ciliary beating so that the cell "backs up" briefly, followed by a resumption of forward motion in a slightly different direction (Figure 8.7). This behavioral response is triggered when the cell bumps into an obstacle or when it encounters a region of altered chemical composition such as an increase in Na^+ concentration. As shown in Figure 8.7, forward swimming is associated with an orientation of the cilia more toward the posterior of the cell, while backward motion involves a shift of ciliary orientation toward the anterior end.

The avoidance reaction in *Paramecium* has been shown to be critically dependent on the presence of Ca^{2+} ions in the medium. The role of Ca^{2+} in this behavioral reaction can be dramatically illustrated in cells that have been killed by treatment with detergents such as Triton X-100, which partially disrupt the plasma membrane. This treatment makes the interior of the cell freely exchangeable to small molecules in the exterior environment. Remarkably, such *Paramecium* "skeletons" still can swim if they are given a source of energy (ATP) and the medium in which they are suspended contains Ca^{2+} ions. If the Ca^{2+}

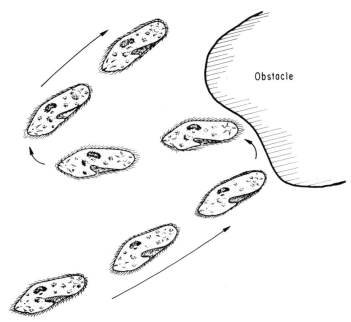

Figure 8.7. The avoidance reaction. When a *Paramecium* cell encounters an obstacle, it reacts by redirecting the beating of its cilia more toward the anterior of the cell. This causes the organism to back up briefly to avoid the obstacle, and then to resume forward motion in a slightly different direction. The arrows denote swimming direction.

concentration is between 10^{-7} and 10^{-8} molar, these skeletons move forward. At about 10^{-6} M Ca^{2+}, however, the extracted cells make little progress in any direction, and above 10^{-6} M Ca^{2+} the skeletons change direction and start swimming backward! Consequently, the direction of the ciliary beat and, therefore, the avoidance reaction itself, can be modulated solely by the Ca^{2+} concentration in the basal region of the cilia.

If the anterior of a *Paramecium* cell encounters an obstacle, the typical avoidance reaction described above occurs. In contrast, if the *posterior* of the cell is mechanically stimulated, for example by a small needle, the organism responds by increasing its rate of forward movement. This "escape reaction" seems not to depend on Ca^{2+}, as the avoidance reaction does, but rather on the ratio of the intracellular to extracellular K^+ concentration. The escape response is depressed as the extracellular level of K^+ is increased.

Paramecium cells are sufficiently large that microelectrodes can be inserted into their cytoplasm. Such impaled cells typically exhibit a resting membrane potential of between -20 and -40 mV, which is primarily due to K^+ and Na^+ permeabilities of the resting membrane (Chapter 6). If the anterior of such cells is mechanically stimulated, or if a small depolarizing current is pulsed into the cytoplasm, an action potential results with depolarization spreading across the entire membrane surface (Figure 8.8). The magnitude of the potential spike is not constant, but depends on factors such as the strength of the stimulus and the extracellular concentration of Ca^{2+}. Nevertheless, membrane potential differences between the peak of depolarization and the resting membrane can be in excess of 50 mV depending on the ionic environment. Within about 50 msec after the initial stimulus, the membrane potential returns to its original resting value (Figure 8.8). These membrane potential changes have been correlated with the direction of ciliary beating in *Paramecium*. Thus, depolarization of the membrane results in a subsequent reversal of ciliary beating ("backing up"), while a return to the resting value triggers resumption of forward swimming behavior.

Membrane depolarization is apparently caused by influx of Ca^{2+} down its electrochemical gradient. Since ciliary reversal depends on local intracellular Ca^{2+} concentrations in excess of 10^{-6} M, it appears that the avoidance reaction in *Paramecium* results from a transient opening of Ca^{2+} channels in the plasma membrane that covers the cilia. Return to the resting potential is accompanied first by the closing of these channels, probably as a result of increased intracellular Ca^{2+} concentration, and second by the flow of K^+ out of the cell. The delayed opening of K^+ channels is analogous to what occurs in nerve cells of higher animals. A Ca^{2+} pump, which is responsible for the normally low intracellular Ca^{2+} levels, then actively pumps out the accumulated Ca^{2+} ions and the cilia resume their beating mode which drives the organism forward. In these respects, the excitable membranes of these protozoa resemble those of typical nerve cells, where Ca^{2+} is the ion responsible for depolarization (Chapter 6).

In contrast to the depolarization of the membrane and the resulting action

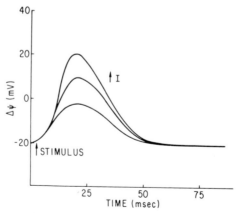

Figure 8.8. Action potentials recorded upon electrical stimulation of a *Paramecium* cell. An intracellular microelectrode is used to introduce a small current pulse into the cytoplasm. Depolarization of the membrane is then recorded as a function of time. Unlike the responses in nerve cells, the magnitude of the response is proportional to the amount of current, I, injected. (Adapted from R. Eckert, Y. Naitoh, and K. Friedman, *J. Exp. Biol.* 56:683–694 (1972).)

potential generated during the avoidance reaction, the escape behavioral response has been shown to be accompanied by a hyperpolarization of the cell membrane. Mechanical stimulation of the posterior end of *Paramecium* increases the permeability of the membrane to K^+ (i.e., opens K^+ channels). Thus, K^+ flows out of the cell, down its chemical gradient, resulting in a more negative membrane electrical potential. The decrease in intracellular K^+ and/or the change in membrane potential may then be responsible for increased ciliary beating in the forward direction, propelling the cell away from the stimulus. The mechanism by which this is accomplished, however, is as yet unclear.

Ion Channels and Behavioral Mutants in *Paramecium*

The observations detailed in the preceding section provide evidence for the localization of mechanically responsive ion channels in different regions of the cell membrane of *Paramecium*. Since an anterior stimulus elicits a response opposite to that resulting from a posterior stimulus (i.e., depolarization versus hyperpolarization), and channels of different specificities are initially activated in each case (Ca^{2+} versus K^+), there must be spatial localization of the two receptor channels at opposite ends of the cell. Furthermore, experiments in which *Paramecium* cells have been deciliated and then allowed to regenerate these structures show that both Ca^{2+} and K^+ channels appear to be localized in the plasma membrane covering the cilia. These ion currents could not be detected in deciliated organisms, but they slowly returned as new cilia were generated. Because the membrane covering the cilia is continuous with the plasma membrane, these observations suggest that lateral diffusion of these channels is restricted in the *Paramecium* membrane.

Both Ca^{2+} and K^+ channels in *Paramecium* exhibit the classic voltage-dependence observed with ion channels in nerve cell membranes. For example, the "open" state of the Ca^{2+} channel is favored by depolarization of the membrane, while at the resting membrane potential, the channels are mostly closed. An intriguing question, however, is how local mechanical perturbation of the cell surface triggers large potential changes across the entire cell membrane. In the case of the avoidance reaction, collision of the anterior end of the cell with an object is believed to cause a local membrane deformation (the nature of which is unknown), which generates a local ion flux called a receptor current and results in a localized depolarization of the membrane. This depolarization is spread via a "cable effect," the entire cell surface is depolarized, and voltage-sensitive Ca^{2+} channels are activated in the cilia. Although the anterior receptor current is probably carried by Ca^{2+} ions, the mechanically activated channels appear to be distinct from the voltage-sensitive ones based on the properties of certain mutant *Paramecium* cells that exhibit altered behavior (see below). Such receptor currents have also been detected in posteriorly stimulated organisms, in which case a local hyperpolarization is the result.

In *Paramecium*, little is known about the molecular characteristics of ion channels that mediate these behavioral responses. Nonetheless, it seems likely that these channels are membrane-spanning proteins that can exist in at least two conformational states by analogy to the nerve cell ion conductors. Modulation of the state of each channel (i.e., open or closed) would presumably be accomplished by conformational changes in channel proteins, which in turn are triggered by changes in the membrane potential and/or mechanical stress.

A promising approach to the study of these behavioral responses in *Paramecium*, and the concomitant ion fluxes associated with them, is the examination of mutant protozoa with altered behavioral properties. Such mutants were first selected on the basis of abnormal responses to chemical stimuli. The largest class of such mutants was termed *pawn*, because cells of this class failed to undergo the avoidance reaction (i.e., they always moved forward like the chess piece of the same name). *Pawn* mutants have been shown to be impaired in their ability to open their voltage-sensitive Ca^{2+} channels. Artificial depolarization of the membrane failed to activate these channels or to reverse the beating mode of the cilia, while injection of Ca^{2+} into these mutant cells triggered the typical avoidance reaction. Thus, the pawn mutations must affect either the Ca^{2+} channel directly or a factor or function upon which the channel depends for its proper operation.

Other types of behavioral alterations have also been observed in various mutant cells of *Paramecium*. *Paranoiac* mutations result in cells that exhibit prolonged episodes of backward swimming owing to persistent depolarization of the membrane in solutions containing Na^+. The defect in these cells appears to relate to an excessive influx of Na^+ during depolarization, an event that tends to increase the length of the action potential, sustaining the depolarization of the membrane for a much longer period of time than normal. In contrast, *fast-2* mutants appear to have abnormalities in K^+ channels such that the resting conductance of the membrane to this ion is much greater than normal, and consequently these cells have a more negative resting membrane potential. *Fast-2* cells fail to generate normal action potentials, exhibit altered avoidance reaction behavior, and swim faster than normal in the forward direction. Biochemical analyses of these and other mutants should provide important clues to the mechanisms controlling behavior in *Paramecium* and other electrically responsive biological systems.

Finally, it should be emphasized that *Paramecium* and related ciliates do not represent isolated instances in which ion currents link sensory reception to a motor response. For example, in the flagellate alga, *Chlamydomonas*, phototaxis appears to involve Ca^{2+} currents generated by a light stimulus, which influence the motion of the flagella. All higher animals also respond to mechano- and chemostimuli with the generation of action potentials, and some evidence exists to suggest that bacteria show rapid fluctuations in membrane potential in response to external stimuli. Bioelectric changes in response to reception processes may prove to be a general mechanism of signal transmission throughout the biological world.

Chemotactic Control of Behavior in *Dictyostelium discoideum*

As discussed in Chapter 4, cellular slime molds including *D. discoideum* proceed from the unicellular state to a multicellular fruiting stage in response to starvation conditions. During vegetative growth individual amoebae seek live bacteria as sources of nutrients, and these are phagocytosed. Detection of food bacteria is accomplished by chemoreception of folic acid and other compounds released by the bacteria. Upon starvation, the amoebae lose sensitivity to folic acid but gain the ability to chemotax toward cyclic AMP. They respond to and secrete pulses of cyclic AMP emitted by a synchronized population of developing cells in a chemotactic process that results in the formation of cohesive cell aggregates (Chapter 10). These aggregates then synthesize an extracellular sheath, and the resultant slug migrates in response to light, warmth, and oxygen. At this developmental stage, the slime mold cells are photo-, thermo-, and aero-tactic. Sensitivity to cyclic AMP is retained during this stage, and the cyclic nucleotide probably influences the progression of events that comprise the developmental sequence. It has been suggested that taxis towards cyclic AMP and oxygen may be important to the process in which pre-stalk and pre-spore cells within the migrating slug sort themselves out (see Chapter 5). Finally, *D. discoideum* cells of opposite mating type are attracted to immature macrocysts, groups of cells where zygote formation occurs, apparently by a chemotactic mechanism. Thus, each phase of the life cycle of *Dictyostelium* is apparently controlled by chemotactic agents.

The chemotactic process that has been most extensively characterized is the aggregation phenomenon which, in *D. discoideum* and several other slime mold species, is directed by cyclic AMP. Some aggregation-competent slime molds are insensitive to cyclic AMP but sensitive to other compounds serving the same function. *D. minutum* and *D. lacteum,* for example, respond to two distinct, species-specific chemoattractants, which may include aromatic or heterocyclic moieties linked to glycine by secondary amide bonds. Slime molds of the genus *Polysphondylium* secrete and respond to other chemotactic agents—oligopeptides that direct aggregation. While the chemoattractant may be species-specific, the aggregation process is universal, and mechanistic features of the process can be expected to be shared by all of the cellular slime molds.

The response reactions of *D. discoideum* amoebae to cyclic AMP are exceptionally complex. The cells, (a) migrate, (b) differentiate, and (c) secrete cyclic AMP in response to extracellularly supplied pulses of cyclic AMP. There is evidence that while one or two cell surface receptors may mediate these processes, the catalytic elements of the biochemical machinery that comprises the transmission pathways differ for each of these three processes. This divergent set of responses is illustrated schematically in Figure 8.9.

Binding of cyclic AMP to the external binding site of its integral membrane receptor elicits three responses. First, pulses of cyclic AMP administered at appropriate intervals stimulate acquisition of aggregation competence. This trait

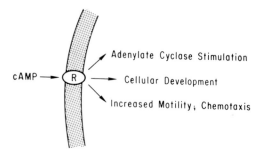

Figure 8.9. Divergent response pathways to the binding of extracellular cyclic AMP to the cell surface cyclic AMP receptor(s) in *Dictyostelium discoideum*. The temporally regulated binding of external cyclic AMP to the cell surface receptor has three independent effects: (1) It stimulates adenylate cyclase, causing a buildup of internal cyclic AMP, which is then released in order to relay the chemotactic signal to other cells; (2) It promotes differentiation of cells from the vegetative state through the fruiting process; and (3) It increases cellular motility locally by increasing the contractile activity of the actomyosin complex, thereby allowing chemotactic orientation.

probably involves synthesis and secretion of the cell surface macromolecules involved in homotypic adhesion (Chapter 10). Interestingly, when the concentration of cyclic AMP is maintained constant, instead of oscillating with a pulse frequency of 6–8 min, development may be inhibited. Cyclic AMP also stimulates subsequent differentiation of cells into pre-stalk and pre-spore cells during slug migration (Chapter 5).

Second, the activity of adenylate cyclase is stimulated in response to cell surface reception of cyclic AMP. This stimulation response is responsible for the buildup of intracellular cyclic AMP (up to 1000-fold over the basal concentration) and for its subsequent secretion into the external medium. Enhanced adenylate cyclase activity is transient, and followed by a refractory period during which the enzyme is insensitive to cyclic AMP-promoted stimulation. This cyclic process of activation–refraction constitutes the amplification element of the chemotactic relay system, which ensures that each cell receiving the cyclic AMP-mediated aggregation signal will pass it on to other cells, more distant from the aggregation center.

Third, pulses of external cyclic AMP, relayed to the population of cells migrating toward the aggregation center, elicit pulses of motion that are easily seen by time-lapse photography. When pulses of cyclic AMP are applied out of phase, the cells will respond by phase shifting, or may be incapable of responding at all, depending on the time when the pulse is administered. These results show that not only adenylate cyclase but also the motility system responds to the binding of cyclic AMP to its receptor in a causal fashion.

Proteins implicated in the chemotactic process are listed in Table 8.1. As noted above, starved slime mold cells alternate between phases in which they produce or respond to cyclic AMP. In response to a rise in extracellular cyclic

Table 8.1. Catalytic Components of the Chemotactic Signal Detection–Relay–Response System in *Dictyostelium discoideum*

Cyclic AMP receptor
Adenylate cyclase
Cyclic AMP transport system
Cell surface and extracellular
  phosphodiesterases
Extracellular cyclic nucleotide
  phosphodiesterase inhibitor
Ca^{2+} channel
Ca^{2+} pumping ATPase
Calmodulin
Guanylate cyclase
Cyclic GMP phosphodiesterase
Myosin heavy chain kinases ⎫ Proposed to be dependent on cyclic GMP and/or the
Myosin light chain kinases ⎭ Ca^{2+}-calmodulin complex.
Myosin heavy and light chains
Actomyosin complex

AMP, adenylate cyclase activity increases dramatically and transiently. Within several minutes, cytoplasmic cyclic AMP concentrations rise as much as 1000-fold. This rise in internal cyclic AMP precedes the efflux of the cyclic nucleotide from the cell. In animal cells, cyclic AMP transport is effected by an energy-requiring transport system that catalyzes extrusion of the nucleotide from the cell. A similar mechanism may be operative in slime molds, but an exocytotic mechanism has not been eliminated.

Upon release from the cells, cyclic AMP diffuses to neighboring cells where it binds to the cell surface receptors. Receptor-binding studies have shown that there are two populations of receptors that differ with respect to their affinities for cyclic AMP. While vegetative cells possess less than 2000 cell surface cyclic AMP-binding sites, aggregation-competent cells possess about 20,000 high-affinity binding sites (K_D = 10 nM) per cell, and about 200,000 low-affinity binding sites (K_D = 150 nM) per cell. Because the low-affinity sites appear first during development, while the high-affinity sites appear later, these two receptors are probably distinct integral membrane proteins.

If the extracellular cyclic AMP were not destroyed, the concentration of this nucleotide would increase and swamp out the spatial and temporal gradients established by secretion. Alternatively, if the cyclic nucleotide were destroyed too rapidly, the maximal concentration in the medium would be insufficient to relay the message from cell to cell. Consequently, the amount of the enzyme that hydrolyzes cyclic AMP, cyclic AMP phosphodiesterase, must be strictly regulated. This has been found to be the case. Extracellular cyclic AMP promotes secretion of the phosphodiesterase, and this enzyme appears in two forms. One form is a fully soluble glycoprotein found in the medium; the other is a similar glycoprotein associated with the external surface of the slime mold membrane.

The first form is subject to inhibition by a glycoprotein (of 45,000 molecular weight) which forms a tight complex with the enzyme. This inhibitory protein is also synthesized and secreted by the slime mold cells, but the rate of its secretion is *inversely* proportional to the extracellular concentration of cyclic AMP. While the inhibitor is not effective on the membrane-bound phosphodiesterase, solubilization of the enzyme unmasks sensitivity to inhibition. Therefore, these two enzymes may be the product of a single gene. These observations lead to the conclusion that the rates of secretion of the phosphodiesterase and its inhibitor maintain the extracellular concentration of cyclic AMP in a range that permits efficient signal propagation through the cell population.

The importance of the correct balance of extracellular phosphodiesterase and its inhibitor to the aggregation phenomenon has been revealed by mutant analyses. Aggregation-negative mutants have been isolated, and the relevant mutants have been found to fall into at least four classes. One class lacks extracellular cyclic AMP phosphodiesterase, while a second produces increased amounts of this enzyme. A third group of mutants cannot synthesize the inhibitor, while a fourth class produces the inhibitor, but production is delayed relative to the aggregation-competent parental strain. Thus, both the hydrolytic enzyme and its inhibitor must be synthesized in the right amounts and at the right times during development in order for the cells to develop normal aggregation competence.

From this discussion it is clear that three principal factors determine the shape of the cyclic AMP pulse: (1) the cyclic AMP receptor-mediated stimulation of adenylate cyclase, (2) the desensitization of adenylate cyclase or adaptation of the enzyme to a refractory state in the presence of external cyclic AMP, and (3) the proper destruction of cyclic AMP by exogenous and cell surface cyclic AMP phosphodiesterase. Other factors may also prove to be important to the maintenance of stable oscillating pulses of cyclic AMP.

The above discussion provides us with insight into the mechanism by which waves of cyclic AMP synthesis are elicited, propagated, and controlled in a population of aggregating slime mold amoebae. It does not, however, allow us to understand how cyclic AMP temporarily promotes increased motility and the chemotactic response. Microscopic observations of *Dictyostelium* in gradients of cyclic AMP have shown that chemotactic movement of this organism is very different from that of bacteria. The amoebae move in approximately straight lines, and the local application of cyclic AMP from a micropipette to any surface of the slime mold cell induces the rapid production of a new advancing front. A new pseudopod forms near the point of administration. Simultaneously, the old front retracts. The gradient of cyclic AMP is apparently sensed locally and may involve a spatial response over the cell surface, or a temporal response during pseudopod extension. In fact, both spatial and temporal response mechanisms may be operative, since the concentration gradient from one end of an amoeba to the other may be no more than 1% of the total concentration under conditions that allow normal aggregation.

The biochemical mechanisms by which an increase in cellular motility results from association of cyclic AMP with its cell surface receptor are just beginning to come to light. Because a change in cell orientation occurs within 5 sec of cyclic AMP stimulation, any biochemical response regulating the motility response must be at least this rapid. Such rapid responses to pulses of external cyclic AMP include (1) an increase in the rate of Ca^{2+} influx as a result of the opening of Ca^{2+} channels in the membrane, and (2) an increase in the cytoplasmic concentration of cyclic GMP, owing to the activation of guanylate cyclase. Both of these responses are rapid and transient as would be required for the mediation of a short-lived signal. A subsequent decrease in the cytoplasmic Ca^{2+} concentration following opening of Ca^{2+} channels would be due to Ca^{2+} extrusion driven by a Ca^{2+} pump, probably a Ca^{2+}-transporting ATPase. The decrease in the cytoplasmic concentration of cyclic GMP following guanylate cyclase activation is attributable to its hydrolysis by a cytoplasmic cyclic GMP phosphodiesterase as well as to its excretion into the external medium. It is of interest that species of slime molds that do not use cyclic AMP as the attractant nevertheless respond to their species-specific attractants with analogous increases in internal Ca^{2+} and cyclic GMP concentrations. No such increase in cyclic AMP concentration is observed in these species. Therefore, cyclic AMP probably does not function in the internal chemotactic transmission signaling that gives rise to a motile response. Instead it may function only extracellularly in those species utilizing cyclic AMP as the aggregation promoting attractant. Probable second messengers therefore include Ca^{2+} and cyclic GMP.

In efforts to piece together a plausible mechanism for regulating the motile response, the protein components of the cellular actomyosin complex, responsible for contractility, and presumably responsible, in part, for cell movement, have been studied. The actomyosin complex in *Dictyostelium* consists of actin and associated regulatory proteins, troponin and tropomyosin, as well as the heavy and light chains of myosin. Regulation of the association and contractility of the actomyosin complex is controlled by phosphorylation, possibly at more than one site. Coincident with the increases in cytoplasmic Ca^{2+} and cyclic GMP concentrations is a rapid decrease in the degree of phosphorylation of the myosin heavy chains. Myosin heavy chain kinase is inhibited by a Ca^{2+}-calmodulin complex. Further, when the myosin heavy chains are phosphorylated, they exhibit lower actin-activated ATPase activity, and polymerization of myosin is inhibited. These observations lead to postulation of the sequence of events depicted in Figure 8.10. Cyclic AMP binding to its receptor opens Ca^{2+} channels and activates guanylate cyclase, possibly by allosteric mechanisms. As a result, cytoplasmic concentrations of Ca^{2+} and cyclic GMP increase. The Ca^{2+}-calmodulin complex binds to myosin heavy chain kinase, while cyclic GMP may bind to a cytoplasmic cyclic GMP-binding protein, a regulatory component of a second protein kinase somehow involved in regulation of the motile response. While the mechanism by which cyclic GMP regulates chemotaxis is not understood, binding of the Ca^{2+}-calmodulin complex to myosin heavy chain kinase inhibits its activity, thereby decreasing the steady state phosphorylation level of myosin. An increase in actomyosin contractility results, owing to enhancement of the myosin ATPase activity and to the tendency of the complex to polymerize. While this description

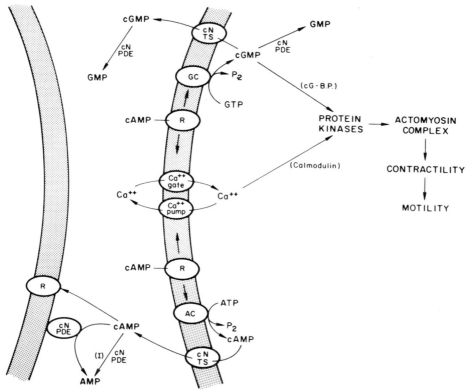

Figure 8.10. Schematic and speculative scheme for the regulation of cellular motility (top) and generation of a cyclic AMP (cAMP) relay system (bottom) as may occur during aggregation of *Dictyostelium discoideum* amoebae. The binding of cAMP to its receptor (R) is thought to open Ca^{2+} channels transiently (Ca^{2+} gate) and activate guanylate cyclase (GC). Cytoplasmic cyclic GMP is hydrolyzed by a cyclic GMP phosphodiesterase (cG PDE) and is excreted via a cyclic nucleotide transport system (cN-TS) into the external medium where it can be hydrolyzed (cN PDE). Cytoplasmic Ca^{2+} binds to calmodulin, and the Ca^{2+}–calmodulin complex inhibits the myosin heavy chain kinase. Cyclic GMP may also alter the activity of a protein kinase in a process mediated by a soluble cyclic GMP binding protein (cG-B.P.). The kinases act to phosphorylate the protein components of the actomyosin complex, which is responsible for contractility and cell motility. The chemotactic relay system (bottom) involves activation of adenylate cyclase (AC) by the cyclic AMP–receptor (R) complex in the membrane. The buildup of cytoplasmic cyclic AMP precedes excretion of the cyclic nucleotide to the extracellular medium, catalyzed by a cyclic nucleotide transport system (cN-TS). The nucleotide diffuses to and elicits a response in a nearby cell analogous to those shown for the first cell. The exogenous cyclic AMP is hydrolyzed by soluble and membrane bound cyclic nucleotide phosphodiesterase (cN PDE). The activity of the former enzyme is controlled by a protein inhibitor (I).

is still incomplete, it is at least possible to envisage a series of causal relationships by which a chemotactic signal at the cell surface can control the motile behavior of the aggregating cell.

Selected References

Berg, H.C. Chemotaxis in bacteria, *Ann. Rev. Biophys. Bioeng. 4:*119 (1975).

Boyd, A. and M. Simon. Bacterial chemotaxis, *Ann. Rev. Physiol. 44:*501 (1982).

Cone, R.A. and J.E. Dowling, eds. *Membrane Transduction Mechanisms,* Raven Press, New York, 1979.

Devreotes, P.N. "Chemotaxis" in *Development of Dictyostelium* (W.F. Loomis, ed.), Academic Press, New York, 1982.

Doetsch, R.N. and R.D. Sjoblad. Flagellar structure and function in eubacteria, *Ann. Rev. Micro. 34:*69 (1980).

Eckert, R. Bioelectric control of ciliary activity, *Science 176:*473 (1972).

Eckert, R. and P. Brehm. Ionic mechanisms of excitation in *Paramecium, Ann. Rev. Biophys. Bioeng. 8:*353 (1979).

Gerisch, G. Chemotaxis in *Dictyostelium, Ann. Rev. Physiol. 44:*535 (1982).

Jacobson, G.R. and M.H. Saier, Jr. "Neurotransmission," in *Biochemistry* (G. Zubay, ed.), Addison-Wesley Publishing Co., Inc., Reading, Massachusetts, 1983.

Korn, E.D. Biochemistry of actomyosin-dependent cell motility (a review), *Proc. Natl. Acad. Sci. USA 75:*588 (1978).

Koshland, D.E., Jr. *Bacterial Chemotaxis as a Model Behavioral System,* Distinguished Lecture Series of the Society of General Physiologists, Vol. 2, Raven Press, New York, 1980.

Koshland, D.E., Jr. Biochemistry of sensing and adaptation in a simple bacterial system. *Ann. Rev. Biochem. 50:*765 (1981).

Kung, C., S.-Y. Chang, Y. Satow, J. van Houten, and H. Hansma. Genetic Dissection of behavior in *Paramecium, Science 188:*898 (1975).

Loomis, W.F. *Dictyostelium discoideum, A Developmental System,* Academic Press, New York, 1975.

Loomis, W.F. Biochemistry of aggregation in *Dictyostelium, Devel. Biol. 70:*1 (1979).

Ross, F.M. and P.C. Newell. Streamers: chemotactic mutants of *Dictyostelium discoideum* with altered cyclic GMP metabolism, *J. Gen. Micro. 127:*339 (1981).

Spudich, J.A. and A. Spudich, "Cell Motility" in *Development of Dictyostelium* (W.F. Loomis, ed.), Academic Press, New York, 1982.

Stock, J.B. and D.E. Koshland, Jr. "A cyclic mechanism for excitation and adaptation" in *Current Topics in Cellular Regulation,* Vol. 18, Academic Press, New York, 1981.

Swanson, J.A. and D.L. Taylor. Local and spatially coordinated movements in *Dictyostelium discoideum* amoebae during chemotaxis, *Cell 28:*225–232 (1982).

CHAPTER 9

Pheromone-Mediated Control of Sexual Activity

Chemical agents regulate cellular metabolic activities, rates of protein synthesis, and spatial morphogenesis in some, and probably all differentiating multicellular biological systems (Chapters 1, 2, 5, and 7). If sexual conjugation evolved as an essential component of developmental programs, permitting species preservation in metazoan organisms in which somatic cells undergo irreversible linear differentiation and programmed death, we might expect that similar agents and mechanisms would have co-evolved to regulate germ cell interactions. Moreover, if analogous mechanisms existed prior to the divergence of the prokaryotic and eukaryotic kingdoms, we can anticipate that chemical control of conjugation may still exist in some present-day prokaryotes. Indeed, Gram-positive bacteria have been reported to release extracellular peptide signals that trigger conjugation. Considerable evidence supports the postulate that chemical agents control sexual activities in a variety of single-celled, differentiating eukaryotes including yeast (*Saccharomyces*), slime molds (*Dictyostelium*), and protozoa (*Tetrahymena*). The same has been amply demonstrated in higher multicellular organisms including Coelenterates (*Hydra*), various insects (*Bombyx, Drosophila*), and humans. In fact, there appear to be no well-documented instances of higher eukaryotes in which extracellular chemical agents (pheromones, hormones, neurotransmitters) do *not* mediate sexual activities. In this chapter we shall consider a few of the better known examples of pheromone control of conjugation. The relevance of chemoreception and electric signal transmission (discussed in Chapter 8), and receptor-mediated regulatory mechanisms eliciting second messenger chemical signal transduction (discussed in Chapter 7), to the mechanisms by which pheromones control sexual activities will become apparent.

Control of Bacterial Conjugation by Sex Pheromones

Bacteria of a single species frequently are capable of conjugal DNA transfer. Conjugation requires the presence of a donor cell (male) as well as a recipient (female). The donor cell usually bears a conjugative plasmid, which carries the genes whose expression is necessary for DNA transfer. A particular bacterial species may possess any one of several different types of conjugative plasmids that code for DNA transfer genes, and possession of any one such plasmid confers upon that cell the capacity to transfer genetic material to an appropriate recipient cell.

Studies with a variety of bacterial species have provided evidence that both the male and female must play active roles in order for conjugal DNA transfer to occur. In a few bacteria, mating between cells of opposite sex is mediated by oligopeptide pheromones produced by the female. The best characterized of such examples is the bacterium *Streptococcus faecalis*.

Recipient cells of *S. faecalis* synthesize and excrete into the extracellular medium a number of diffusible sex pheromones which induce donor cells harboring appropriate conjugative plasmids to become adherent to the female cells (Figure 9.1). This induction process has been shown to require transcription and translation of male-specific genes and can be elicited by exposure of the male cells to a partially purified preparation of the active pheromone. While these sex-specific compounds exhibit *clumping-inducing* activity, they also enhance the frequency of mating. Thus, in one instance, exposure of a plasmid-bearing donor cell to the cell-free culture medium derived from a plasmid-free recipient enhanced the rate of subsequent plasmid transfer more than a millionfold when the recipient was added to the preconditioned donor.

Recipient cells actually produce several distinct sex pheromones, each specific for a different class of donors bearing plasmids. During conjugation, receipt of a particular conjugative plasmid by the recipient results in rapid cessation of the secretion of the active pheromone specific for that plasmid. Some evidence suggests that in at least some instances, conjugative plasmids carry a gene (or genes) that inactivates the sex pheromone controlling its own transfer. Interestingly, however, other pheromones specific for donors bearing other classes of plasmids are synthesized unabated. All of the sex pheromones so far examined are peptides with molecular weights of less than 1000. They are heat-resistant but are inactivated by proteolytic enzymes.

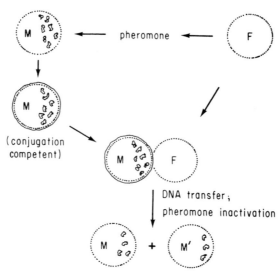

Figure 9.1. Control of plasmid-directed conjugation in *Streptococcus faecalis* by sex pheromones. The recipient cell (F) produces a peptide pheromone that diffuses into the medium and acts upon the donor cell (M) to induce clumping activity as well as conjugation competence. Plasmid transfer results in conversion of the recipient cell to a new donor (M'), and the pheromone produced by that cell is inactivated.

Recently it has been shown that the sex pheromones of *S. faecalis* not only modify the donor cell surface such that sexual agglutination is promoted, but they also induce other donor-specific conjugation functions. In these experiments, mating was studied between two donor strains bearing conjugative plasmids of different classes. Each donor cell served as the recipient for the other. Since exposure of one donor to the sex pheromone of the other donor fostered rapid cell clumping, bidirectional conjugative transfer was expected to occur if cell aggregation alone was induced by pheromone exposure. However, it was found that exposure of each donor to the pheromone of the other donor promoted unidirectional transfer. Bidirectional transfer occurred only if both cell types had been exposed to the pheromone produced by the other cell. It can therefore be concluded that multiple sex-specific functions are induced when a donor bacterium is exposed to the pheromone produced by an appropriate recipient cell. The mechanisms by which bacterial peptide pheromones enhance conjugation-competence have yet to be elucidated.

Pheromone Control of Yeast Mating

Within any one species, mating of haploid yeast cells of opposite mating type (α and **a**) is controlled by two oligopeptide pheromones, the α-factor (generated by haploid α cells and specifically acting on **a** cells), and the **a**-factor (produced by **a** cells and acting on α cells). The multifaceted function of these pheromones is to permit vegetative cells to differentiate into cellular forms possessing the essential characteristics of gametes and to induce the functions that mediate fusion of gametes of opposite mating type. The pathway by which α-factor and **a**-factor promote mating in *S. cerevisiae* is depicted schematically in Figure 9.2.

α-Factor acts on **a** cells to arrest growth in the early part of the G1 phase of the cell cycle, prior to initiation of budding. Similarly, **a**-factor arrests α cells in the early G1 phase of the cell cycle. In each cell type, initiation of DNA synthesis is blocked within 2 hr after addition of the pheromone. Three to 4 hr after pheromone addition, the haploid yeast cells begin to undergo morphological changes. They elongate from one point in the spherical cell, producing a pear-shaped cell or "shmoo" in a process morphologically reminiscent of early germ tube formation in *Neurospora* (Chapter 2). In the electron microscope the elongated tips of the pear-shaped α and **a** haploid cells exhibit a thinner wall layer with a fuzzier, extracellular coat as compared with the remainder of the cell wall (see Figure 10.2 in Chapter 10). There is probably somewhat more glucan and chitin in the tip region, and different mannoproteins are synthesized in response to pheromone exposure. Among the macromolecular constituents of the tip walls are the complementary sexual agglutination factors that mediate heterotypic intercellular adhesion (see Chapter 10).

Sexual agglutination triggers local changes in the cell wall structures of both

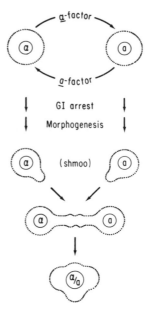

Figure 9.2. Schematic depiction of events giving rise to the fusion of haploid yeast cells of opposite mating type. This process is analogous to fertilization of a haploid egg cell by a sperm cell in higher organisms (see Chapter 10).

haploid cells. The rigid glucan layers become progressively thinner, probably as a result of the autolytic activities of cell wall glucanases. The membranes of the adjacent cells then come into near contact and fragment by a vesiculation process that allows cytoplasmic continuity of the apposed haploid partners. Cell fusion gives rise to a heterokaryon (a cell with two nuclei), and this fusional event apparently prevents fusion of other haploid cells to the binucleate cell, possibly by an active exclusion process analogous to the "block to polyspermy" observed upon fertilization of haploid animal egg cells (Chapter 10). Consequently, the frequency with which three or more haploid yeast cells fuse is less than one-ten-thousandth of the frequency of normal diploid mating, even under conditions allowing for mass agglutination of the haploid cells. Finally, the two nuclei of opposite mating type fuse to give a true diploid cell with a single membrane-enclosed nucleus of the $\alpha/\mathbf{a}$ genotype.

The oligopeptide pheromone, α-factor, produced by α cells, contains 13 amino acyl residues of known sequence as shown in Figure 9.3A. The analogous $\mathbf{a}$-factor is a mixture of two related oligopeptides, each containing 11 amino acyl residues and differing only at a single position. Their sequences are shown in Figure 9.3B. Although detailed mechanistic studies have been carried out only with α factor, the two agents appear to act on the haploid cell of opposite mating type in a strictly analogous fashion. Recent studies have shown that $\mathbf{a}$ cells, but not α cells or $\mathbf{a}/\alpha$ diploids, possess on their surfaces a receptor for α-factor that binds the pheromone with a dissociation constant of about 10^{-7} M. There are estimated to be approximately 10^5 receptors per cell. Moreover, $\mathbf{a}$ cells can destroy α-factor, allowing them to recover from growth arrest. Destruction of α-factor activity results from proteolytic cleavage by a protease specifically associated with the $\mathbf{a}$-cell surface. Since extracts or spheroplasts of α-cells also hydrolyze α-factor (although intact α cells do not), differences in access, not presence of the protease(s), may be responsible. The cleaved inactive α-factor fragments are quantitatively recovered in the culture medium, even after prolonged incubation with $\mathbf{a}$ cells. This observation eliminates the possibility that the mechanism of pheromone action involves internalization of a proteolytically clipped fragment.

α-Factor has been shown to markedly inhibit the cyclic AMP synthetic enzyme, adenylate cyclase, in plasma membrane fragments isolated from $\mathbf{a}$ cells. By

(A) H_2N-Trp-His-Trp-Leu-Gln-Leu-Lys-Pro-Gly-Gln-Pro-Met-Tyr-CO_2H.

(B) H_2N-Tyr-Ile-Ile-Lys-Gly-Val-Phe-Trp-Asp-Pro-Ala-CO_2H
      H_2N-Tyr-Ile-Ile-Lys-Gly-Leu-Phe-Trp-Asp-Pro-Ala-CO_2H

Figure 9.3. Amino acyl sequences of the yeast peptide pheromone, α-factor (shown in A) and the two $\mathbf{a}$-factors of nearly identical sequences (shown in B). α-Factor is produced by α haploid cells and acts on $\mathbf{a}$ cells while the two $\mathbf{a}$-factors are produced in $\mathbf{a}$ cells and act on α cells. Pheromone reception appears to promote mating by lowering the cellular concentration of cyclic AMP (see text).

contrast, when the enzyme is solubilized with the detergent, Triton X-100, it no longer exhibits sensitivity to pheromone inhibition. Moreover, only biologically active pheromone analogues inhibit, and there is a quantitative correlation between biological activity (measured with intact cells) and inhibition of adenylate cyclase (measured *in vitro*). These results lead to the suggestion that α-factor binding to its receptor on the external surface of the **a** cell membrane exerts its biological effects by reducing the intracellular concentration of cyclic AMP.

This *second messenger* hypothesis (see Chapter 7) is further substantiated by two additional observations. First, when cyclic AMP (1 to 10 mM) is added to **a** cell suspensions after treatment with α-factor, the cyclic nucleotide accelerates recovery from α-factor-induced G1 arrest. Second, one class of temperature-sensitive **a** cell mutants, selected on the basis of resistance to α-factor-induced growth stasis at 34°C but not at 23°C, exhibited adenylate cyclase activity that was sensitive to α-factor regulation only when grown at the lower temperature. These results lead to the suggestion that the pheromone exerts its effect, at least on G1 arrest, by inhibiting adenylate cyclase activity and thereby lowering the intracellular cyclic AMP concentration. Since many yeast proteins (ribosomal proteins, RNA polymerase subunits, and NAD-dependent glutamate dehydrogenase among others) are reversibly phosphorylated by protein kinases, and since cyclic AMP binding proteins, identified as regulatory components of protein kinases, have been detected in yeast, it appears reasonable to suggest that cyclic AMP exerts its biological effects by controlling the levels of phosphorylation of proteins involved in cell cycle regulation and shmoo formation. Thus, the chain of events leading to the biological response can be visualized as follows: (1) α-factor binds to its receptor on the **a** cell surface. (2) The α-factor-receptor complex inhibits adenylate cyclase by an unknown mechanism that causes a decrease in the cytoplasmic concentration of cyclic AMP. (3) The cyclic AMP-dependent protein kinase therefore shows diminished activity, reducing the degree of phosphorylation of structural or enzymatic constituents of the cell that must be phosphorylated to allow progression through the G1 stage of the cell cycle. (4) Cell cycle arrest results. (5) Concomitantly, other yeast proteins that must be present in the dephosphorylated state in order to promote progression through the mating program are activated due to protein-phosphate phosphatase action under conditions of protein kinase inhibition.

A number of studies have led to significant conclusions regarding the mode of α-factor biosynthesis. This pheromone is apparently synthesized and secreted constitutively in α cells. Since it has been found to consist exclusively of natural L-amino acids, and since cycloheximide-treated cells and RNA and protein biosynthetic mutants are blocked in its synthesis, it has been concluded that the biosynthetic process occurs by a transcription–translation coupled mechanism. Moreover, larger precursors of α factor have been isolated from certain α cell mutants. These α-factor precursors have four additional amino acids at the N-terminus, which are cleaved by a specific proteolytic processing enzyme.

Recently the α-factor structural gene has been cloned using recombinant DNA technology. The nucleotide sequences of the cloned DNA confirm that α-factor is first synthesized as a long precursor polypeptide of 165 amino acyl residues in which four exact copies of the mature α-factor are linked in tandem, separated by spacer amino acid residues, at the carboxyl terminal half of the precursor. The amino terminal half of the presumed precursor consists of a hydrophobic *signal* sequence of about 20 residues followed by a number of hydrophilic residues. Thus, to generate α-factor, the precursor must be processed by precise proteolytic cleavage events involving removal of the amino terminal part of the precursor, excision of the spacer amino acids, and release of the processed α-pheromone.

If the structural gene for **a**-factor were similar to that for α-factor, then it also might encode tandem repeats of the mature **a**-factor sequence separated by spacer amino acid residues. If two of these tandem DNA repeats were to differ by only a single nucleotide substitution, it would be possible to explain the occurrence of the two forms of **a**-factor, which differ in a single amino acid substitution (val for leu). Contrary to this prediction, however, recent experiments with cloned DNA fragments encoding the two **a**-factors have revealed two discrete chromosomal segments, each encoding a larger precursor polypeptide containing only a single copy of the **a**-factor at its carboxyl-terminus. One of these genes encodes the valine-containing **a**-factor, while the other encodes the leucine-containing **a**-factor. This observation emphasizes the versatility with which living organisms have evolved comparable processes of gene expression.

Pheromone Control of Sexual Activities in Insects

Pheromones frequently serve to integrate the behavior of organisms or of populations of individuals within a particular animal species. For example, ants use chemical cues to mark trails to food sources, whereas vertebrates and invertebrates alike employ olfactory signals to locate their mates. Much of the work conducted with animal pheromones deals with those found in insects because of the economic importance of many of these species and the potential for controlling insect populations by using these agents. Insect pheromone systems are among the most completely described systems of chemical communication with respect to structure, mechanisms of reception, and mode of action.

One of the most carefully studied pheromone-chemoreceptor systems is that found in the silk moth, *Bombyx mori*. Bombykol, an unsaturated fatty alcohol (Figure 9.4), is a sex attractant produced by the female and detected by chemoreceptors on the antennae of the male. Of the thousands of antennal receptors available, about one-half respond specifically to bombykol. Although the female's antennae are insensitive to this chemical, those of the male are so sensitive that adsorption of *one molecule of the pheromone* to the receptor surface will elicit a nerve impulse. When a male detects the volatile bombykol molecule, he proceeds upwind unless the scent is lost, in which case he initiates random motion. When the concentration of bombykol reaches a certain critical value,

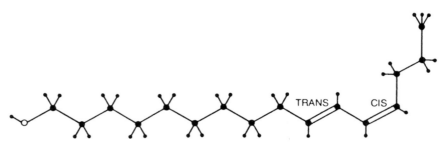

Figure 9.4. Structure of bombykol, the sex pheromone produced by the female silk moth *Bombyx mori*. The molecule is an unsaturated fatty alcohol with the chemical designation *trans*-10-*cis*-12-hexadecandien-1-ol. ● Carbon. ○ Oxygen. • Hydrogen.

behavioral changes can be observed, and the male begins to track the female by an odor gradient.

The chemoreceptors involved are located in antennal bristles, the spacing and diameter of which create a molecular sieve. Adsorption of a molecule depends upon the aerodynamic conditions, the temperature, the concentration of the pheromone, and the physicochemical properties of the molecule and the bristle. The preferential adsorption of bombykol to chemoreceptor hairs is indicated by studies with radioactively labeled bombykol in which 80% or more of the label was found on the chemoreceptor hairs, even though the hairs constitute only 13% of the antennal surface area.

Once adsorbed, the pheromone must penetrate the bristle to the sensory nerves enclosed within the shaft of the bristle. The outer layer of the bristle is formed of cuticle, a hardened matrix of chitin, protein, and lipid, which serves as the exoskeleton of the animal. Bombykol molecules do not penetrate the cuticle but probably pass through pores found along the length of the shaft. Electron micrographs of sections of the antennal hairs show that inside the pores are tiny fluid-surrounded tubules, about 30 Å in diameter, which lead to the sensory nerves. The bombykol molecules presumably penetrate the pores and pass along the tubules or within the liquor to the sensory nerve surface. By testing the effect of other organic compounds, it has been determined that lipid solubility is an important property such that increasing carbon chain length decreases the threshold concentration for stimulation. Thus, the factors affecting diffusion of the stimulating molecules to the nerve account in part for the specificity of the chemoreceptor.

Because radioactively labeled bombykol has never been detected by autoradiography inside the nerve, activation of the nerve is presumed to occur at the membrane surface by means of interactions with a receptor molecule associated with the membrane. From structure–function studies of the stereochemistry of bombykol, it appears that the terminal hydroxyl group and the configuration resulting from the *cis* and *trans* double bonds are vital to the biological activity of the pheromone. Binding to the receptor is thought to elicit a generator potential that induces an action potential near the cell body of the nerve. Although one molecule of bombykol can elicit a nerve impulse, several hundred molecules are required to effect the behavioral response. The nature of the complex integrative process in the brain which produces the tracking response is completely unknown.

Genetic Dissection of Courtship Behavior in *Drosophila*

In the previous sections of this chapter, it was shown that pheromones mediate mating of some prokaryotic and numerous eukaryotic organisms. In higher organisms, regulation of mating may be exceptionally complex, being influenced by visual, tactile, auditory, and olfactory cues. Genetic techniques have been used traditionally in microorganisms to dissect complex processes and gain insight into the molecular mechanisms. The same experimental approaches are at least theoretically feasible in higher organisms, and they are applicable to behavioral studies. In fact, the microbiological approach has already been applied to the analysis of mating behavior in fruit flies of the genus, *Drosophila*. Application of genetic techniques to reproductive behavioral studies presupposes that behavior is genetically programmed by developmentally regulated genes. The genetic evidence for this supposition and the mechanistic implications of mutant analyses of courtship rituals will be considered in this section.

Proper mating responses in some species of *Drosophila* depend primarily on visual cues. By contrast, in other species, such as *D. melanogaster* and *D. subobscura*, visual and olfactory cues are of approximately equal importance. If male and female flies are brought together in the dark, if the male is blind due to a genetic defect, or if the female is genetically paralyzed so that she is incapable of performing her courtship dance, courtship vigor and the copulation rate decline to about half of their normal values. The visual response of the male apparently depends largely on the movements of the female rather than on her appearance.

In the absence of visual cues, mating responses are under pheromone control. The volatile sex pheromones are produced within the abdomen of the female. This was shown in *Drosophila* with male–female sex mosaic flies in which different parts of the organism possessed cells of one or the other sex. Standard courtship behavior with normal male flies was only observed when they were exposed to mosaic flies possessing abdomens that were at least partially female.

Mutant males that cannot detect olfactory cues, termed *smell blind,* are incapable of responding to female sex pheromones, and these males exhibit twofold decreased courtship vigor. Doubly mutant males, which are both smell blind and blind are sexually inactive. These observations reveal the relative significance of visual and olfactory cues in *Drosophila melanogaster*.

Ordinarily, two mature male *Drosophila* do not court each other, but in the presence of the volatile pheromone from the female, homosexual courtship occurs. Moreover, young males, less than one day old, stimulate courtship behavior in older males almost as effectively as do virgin females. The young males have not yet acquired the ability to court other females or males, but they produce a volatile sex pheromone which, like that of the female, can be extracted and purified. While the properties of the pheromones produced by females and young males appear to be similar, the structures of these molecules have yet to be elucidated.

Mutations within a gene on the third *Drosophila* chromosome, termed *fruitless* (*fru*), cause males of any age to resemble young wild type males with regard to much of their sexual behavior. They do not mate with females, but they both court other males and stimulate the other males to court them. Stimulation is passive, requiring no motion or visual cues, and is attributed to the production of an active substance not present in normal mature male or female flies. The active ingredient may be the same as that produced by young males. Thus, *fru* mutations may abolish the process that normally turns off the production of the male-specific pheromone at day one when sexual maturation normally occurs. Possibly, the active pheromone plays three determinative roles in sexual behavior. First, it stimulates a *fru* mutant male to court other males; second, it stimulates other males to court him; and third, it blocks the mutant's response to females.

The normal physiological function of the male pheromone is not entirely clear. However, male *Drosophila* produce a "courtship song" by means of wing movements. It is possible that this song is a "learned" response, learned by imitation of the song of a mature male of the same species. Pheromone production in the normal sexually immature male may therefore function to attract male musical "pedagogues" who teach the young flies the courtship song and behavior of their species. Alternatively, pheromone production may serve as a protective device to prevent aggressive behavior by older males.

Quantitative studies of mating behavior have revealed that mated females of *Drosophila* are less attractive to males than are virgin females. This response, which lasts for 4–7 days, does not depend on male vision or female movements as shown by mutant analyses. However, smell-blind males cannot distinguish between virgin and mated females. These observations suggest that mated females release volatile pheromones that differ qualitatively or quantitatively from those of the virgin fly, and that males distinguish the two by olfaction.

Males exposed to mated females for more than 20 min exhibit an "after effect". They show low courtship vigor when subsequently exposed to a virgin female. This after effect usually lasts 2–3 hr. Initially it was assumed that the effect was due to the production of anti-pheromones or to inactivation of the male olfactory receptors. However, male memory-deficient mutants (amnesiacs), which forget more easily than do wild type flies, exhibit abnormally brief after effects subsequent to exposure to a mated female. They court virgin females with maximal vigor within 15–30 min.

The mechanisms by which males sense mated females and learn to exhibit the after effect are almost certainly mediated by olfactory cues. Immobilized mated females elicit both responses, and the after effect is observed in the dark, or with blind males. Possibly males learn to associate an adverse odor or a mating inhibitory pheromone with the presence of a female. Only when this negative stimulus or the association between the stimulus and a female is forgotten is courtship vigor restored.

The experiments described in this section illustrate how microbiological principles and genetic techniques can yield useful information concerning complex behavioral patterns in higher organisms. As these techniques are refined and our appreciation of their value increases we should be able to learn more and more about the complex molecular interactions governing sex, differentiation, and programmed death in multicellular organisms. While evolution tends towards complexity, technological advances allow us to unravel the complex. An increasing number of organisms should become amenable to the scientific approaches of the microbiologist in the near future.

Selected References

Clewell, D.B. Plasmids, drug resistance, and gene transfer in the genus *Streptococcus*, *Microbiol. Rev. 45:*409 (1981).

Flegel, T.W. The pheromonal control of mating in yeast and its phylogenetic implication: A review, *Can. J. Microbiol. 27:*373 (1981).

Glass, R.E. *Gene Function: E. coli and its Heritable Elements,* University of California Press, Berkeley, 1982.

Goodenough, U.W. and J. Thorner. "Sexual Differentiation and Mating Strategies in the Yeast *Saccharomyces* and in the Green Alga *Chlamydomonas*", in *Cell Interactions and Development: Molecular Mechanisms* (K. Yamada, ed.), John Wiley & Sons, Inc., New York, 1982.

Hall, J.C., L. Tompkins, C.P. Kyriacou, R.W. Siegel, F. von Schilcher, and R.J. Greenspan. "Higher Behavior in *Drosophila* Analyzed with Mutations that Disrupt the Structure and Function of the Nervous System" in *Development and Neurobiology of Drosophila* (O. Siddiqi, P. Babu, L.M. Hall, and J.C. Hall, eds.), Plenum Press, New York, 1980.

Leonard, J.E. and L. Ehrman. Pheromones as a means of genetic control of behavior, *Ann. Rev. Gen. 8:*179 (1974).

Matsumoto, K., I. Uno, Y. Oshima, and T. Ishikawa. Isolation and characterization of yeast mutants deficient in adenylate cyclase and cAMP-dependent protein kinase, *Proc. Natl. Acad. Sci. USA 79:*2355 (1982).

Schneider, D. The sex-attractant receptor of moths, *Scient. Am. 231:*28 (1974).

Sprague, G.F., Jr., L.C. Blair, and J. Thorner. Cell interactions and regulation of cell type in the yeast, *Saccharomyces cerevisiae, Ann. Rev. Micro. 37:*623–660 (1983).

Thorner, J. "Intercellular Interactions of the Yeast *Saccharomyces cerevisiae*" in *Microbial Differentiation* (T. Leighton and W.F Loomis, eds.), Academic Press, New York, 1980.

Thorner, J. "Pheromonal regulation of development in *Saccharomyces cerevisiae*" in *Molecular Biology of the Yeast, Saccharomyces, Vol. I: Life Cycle and Inheritance* (J. Strathern, E.W. Jones, J.R. Broach, eds.), Cold Spring Harbor Laboratories, Cold Spring Harbor, New York, 1981.

Willetts, N. and R. Skurray. The conjugation system of F-like plasmids, *Ann. Rev. Gen.* *14:*41 (1980).

CHAPTER 10

Cellular Recognition: Mechanisms and Consequences of Homotypic and Heterotypic Adhesions

You may object that it is not a trial at all; you are quite right, for it is only a trial if I recognize it as such.

Franz Kafka

It is widely accepted that cell surface macromolecules must mediate recognition phenomena that allow like cells within a tissue to adhere to one another (homotypic adhesion) as well as to cells of other types (heterotypic adhesion). These macromolecular interactions occur by multistep processes that are important in the formation and maintenance of tissues both in the adult organism and during embryonic development. Moreover, we now know that specific cell–cell interactions regulate many physiological processes. Contact inhibition of growth is a well-documented example of negative growth regulation induced by homotypic cellular adhesions. Cellular motility may also be regulated in a negative sense as observed in the phenomenon of contact inhibition of motion. And highly specific intercellular adhesions can induce the syntheses of enzymes and proteins responsible for the expression of tissue-specific traits.

In this chapter we shall first consider cell–cell interactions as a dynamic force during embryonic development and show that when these recognition phenomena are disrupted, embryonic development frequently cannot proceed. Two microbial systems, one involving homotypic adhesions (slime mold aggregation) and the other involving heterotypic adhesions (sexual agglutination in yeast), will be discussed both from functional and mechanistic standpoints. These two systems are probably the best characterized microbial cell adhesion systems presently available. It will then be shown that the interactions of viruses with their host cells are the consequences of highly specific heterotypic adhesion processes that probably have the potential for more detailed genetic and mechanistic analysis than any other type of system. In the subsequent sections of the chapter, some of the biological consequences of intercellular interactions (induction of protein synthesis, contact inhibition of growth, regulation of cellular motility) will be considered. Finally, in the last section we shall examine the process by which sperm–egg contact initiates the program leading to embryological development of the fertilized egg.

t Loci in Mice: Components of a Genetic Program for Embryonic Development

Most adult mouse cells possess on their surfaces the H2 histocompatibility antigens as well as other antigens characteristic of the mature state. By contrast, early embryos, germ cells, and embryonal carcinomas (see Chapter 3) lack the H2 histocompatibility antigens. A principal antigen on many of these latter cells is designated F-9 and may be the product of one gene within the complex *t* locus, the mutant allele of which is called t^{12}. Heterozygotes for the t^{12} allele possess half as much of the F-9 antigen as do cells of the wild type organism. The homozygous *t* state (t/t) is lethal at a particular time in the developmental program. For example, the t^{12} allele, when homozygous, prevents blastocyst formation. Development stops at the morula stage. The t^0 allele, on the other hand, acts at a somewhat later stage in development, preventing division of the inner cell mass which gives rise to embryonic and extraembryonic ectoderm. The homozygous t^9 allele, acting still later, prevents the proper development of intercellular adhesions in mesodermal tissues. Intercellular junctions do not form, abnormal cell shape results, and death follows. Finally, the t^{w1} allele acts at a still more advanced developmental stage, preventing growth and maintenance of the neural tube and brain. Figure 10.1 depicts the consequences of the presumed sequence of inductive and repressive events that allow for selective *t* gene activation and inactivation during embryonic development in mice.

All of the *t* genes map within a well characterized region of mouse chromosome

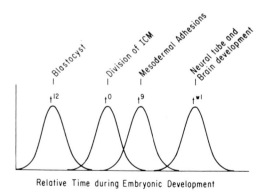

Relative Time during Embryonic Development

Figure 10.1. Proposed sequential expression of the *t* alleles that function to regulate specific steps in the development of the embryonic mouse. The *t* alleles may code for cell surface macromolecules that mediate cell–cell recognition. ICM, inner cell mass.

number 17, not far from the genetic region that codes for the adult histocompatibility antigens. Moreover, the F-9 antigen is a β-microglobulin as is the H-2 histocompatibility antigen. These observations lead to the suggestion that a region of the ancestral mammalian genome duplicated and then diverged. One of the duplicated segments (the *t* region) came to control intercellular recognition during embryonic development. The other became the genetic region coding for surface antigens controlling cell–cell interactions in the adult organism. Recognition of the essential functions of the *t* loci during embryonic development serves to emphasize the possible importance of cell adhesion–recognition phenomena to developmental processes.

Cellular Recognition and Tissue Formation

To move from the widely accepted statement about the importance of cell surface macromolecules in cell adhesion during tissue formation, noted above, to the actual identification of the cell surface macromolecules involved has been extremely difficult. From the point of view of cell–cell adhesion, cell surface macromolecules fall into four categories: molecules that directly mediate cell–cell adhesion (adhesion molecules), molecules that serve as "flags" identifying specific differentiating cell types, but that may not directly participate in adhesion (recognition molecules), molecules that might occur preferentially on the surfaces of certain cell types because of their differentiated function (differentiation molecules), and molecules that function at the cell surface in processes necessary for the survival of all cell types (housekeeping molecules). Clearly these categories overlap, and the critical test is the function of the molecule during normal histogenesis.

One approach to identifying potential adhesion or recognition molecules is to look for stages in histogenesis where new types of cell–cell interaction are required of particular embryonic cells. Cell surface macromolecules that are newly expressed, cease to be expressed, or are somehow rearranged at the cell surface at these stages of histogenesis can then be examined by a variety of *in vitro* assays for their ability to mediate cell–cell adhesion or recognition. The use of several assessment methods is critical since there is no inherently obvious *in vitro* model for the complex variety of *in vivo* cell interactions. Alternatively, cell surface macromolecules initially identified by such assays can be investigated at the critical stages of histogenesis and their role in that particular cell–cell interaction assessed.

Two *in vivo* histogenetic processes that have one or more stages where critical changes in cell interaction, and therefore in cell surface biochemistry, must occur are neurogenesis and skeletal myogenesis. Neurons must form specific transient interactions with other neurons and with glial cells, and eventually form highly specific synapses with other neurons. Skeletal myoblasts must cease dividing, align and fuse with other myoblasts at the proper developmental stage while taking care not to fuse with adjacent fibroblasts and endothelial cells.

Two different cell surface macromolecules have been identified as cell adhesion and/or cell recognition molecules in neurogenesis, although their mechanisms of action *in vitro* are not known. These macromolecules are the neural cell adhesion molecule (N-CAM) and the retina cognin. Both were isolated from the neural retina of chick embryos.

N-CAM is a 140-kilodalton glycoprotein associated with the surfaces of embryonic neural cells from various portions of the nervous system. It is able to mediate the adhesion of nerve cells to each other, in preference to adhesion to other cell types in several assay systems. Antibody prepared to N-CAM perturbs not only cell–cell aggregation, but also the pattern of *in vitro* histogenesis, the fasiculation of neurites, and their response to nerve growth factor *in vitro*. Synthesis of N-CAM, which appears to be antigenically similar to another cell surface protein found in rat brains (the D-2 protein), appears to be abnormally regulated in the mouse neurological mutant *staggerer*. This is particularly interesting since staggerer mutants exhibit defective synaptogenesis.

The retinal cell-aggregating glycoprotein, cognin, is smaller than N-CAM, having a size of 50 kilodaltons, and is less widespread. It is present on the surfaces of retinal neurons in larger amounts than N-CAM. Cognin is an integral membrane protein that enhances the initial adhesion and the subsequent *in vitro* histogenesis of retinocytes from various vertebrate embryos. It has no effect *in vitro* on neurons from other regions of the brain because it does not bind to these cells. Antibodies directed against cognin block retinocyte cell adhesion and the pattern of histogenesis *in vitro*. The distribution of cognin on the surface of retinocytes *in vivo* changes with neurogenesis from a rather uniform distribution on virtually all neurons in the retina to a patchy distribution, where it occurs on only parts of the surface of a minority of the retinal neurons. Some observations have led to the suggestion that cognin may play a role in synapse formation.

Skeletal myogenesis has long been studied *in vitro,* and myoblasts from a variety of embryonic sources pursue a predictable pattern of initial mitosis, migration and alignment (cell recognition), and subsequent cell type-specific fusion. Cell surface changes have clearly been implicated in many of these steps, and critical individual cell surface macromolecules are now being identified. Here, however, their cell-adhesion properties are not yet known. Perturbations in the cell surface transferrin receptor correlate with the cessation of mitosis, and the appearance of a receptor for prostaglandin E_1 in myoblasts correlates with the cell–cell recognition step. In these cases the consequences of the changes in cell adhesion are known, and the correlation with modification in critical candidate cell surface macromolecules has been established. As in neurogenesis, the mechanisms of cell interaction need to be determined.

Sexual Agglutination in Yeast

As discussed in Chapter 9, pheromone action promotes the syntheses of cell surface macromolecules that mediate the sexual agglutination of haploid yeast strains of opposite mating type. α Cells of *Saccharomyces cerevisiae* adhere specifically to **a** cells but not to other α cells or to **a**/α diploids. The adhesive molecules are apparently localized to the elongated tips of the pheromone-induced cellular extensions of the *shmoos*. Electron microscopy has revealed that only in the region of elongation is the cell surface thick and fuzzy, and that membrane vesicles, possibly carrying sexual agglutinins and other cell wall and membrane macromolecules to the surface of the growing tips, underlie the plasma membrane in this region only (Figure 10.2). It would be reasonable to hypothesize that

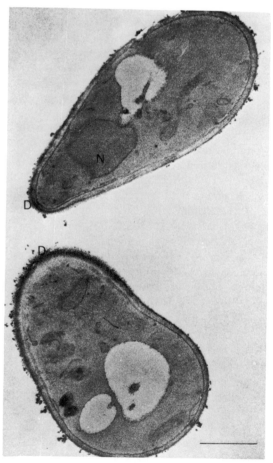

Figure 10.2. Thin section electron micrograph of a developing *S. cerevisiae* "shmoo." Pheromone-induced morphogenesis of the elongated cell from a spherical one involves insertion of cell surface macromolecules, which give the tip a very different appearance from the remainder of the cell. Membrane vesicles that underlie the elongated tip probably bring sexual agglutinins and other macromolecules to the surface and function in the biosynthesis of the "gametic" cell wall. In this case **a** cells were treated with α-factor for 3 hr before fixation. From P.N. Lipke, A. Taylor, and C.E. Ballou, *J. Bact. 127*:610–618 (1976), copyright American Society for Microbiology, reprinted with permission.

ionic currents and transcellular electric fields direct the synthesis of the growing shmoo tips by causing electrophoresis of biosynthetic vesicles. The transcellular electric field might be generated in response to asymmetric pheromone binding (receptor clumping). A mechanistic analogy is thus suggested for germination in the brown algae, *Fucus* and *Pelvetia* (Chapter 2), germ tube generation in *Neurospora* (Chapter 4), and pheromone-induced sexual elongation of haploid yeast cells of opposite mating type (Chapter 9).

While synthesis and expression of sexual agglutination factors are normally induced in response to pheromone binding in most yeast strains, other strains synthesize these macromolecules constitutively. In all yeast strains, synthesis of these molecules is apparently controlled by regulatory genes encoded at the mating type locus (MAT) (see Chapter 12). For example, genetic evidence suggests that a negative control element, or repressor, regulates inducible expression of the sexual agglutinin on the **a** cell surface. Induction of the agglutinin occurs when **a** cells are exposed to α-factor. A defect in the gene encoding the repressor can give rise to a constitutive phenotype, and this defect is recessive to the wild type allele, permitting inducible synthesis of the agglutinin. A mutation that gives rise to the constitutive phenotype has been mapped to the *MAT***a** locus, which suggests that a protein encoded within *MAT***a**, perhaps the pa2 protein, is the repressor of sexual agglutinin synthesis (Chapter 12).

Isolation of the yeast cell surface macromolecules that mediate sexual agglutination was first carried out with the yeast species *Hansenula wingei*. Haploid vegetative cells of one mating type (called 5-cells) agglutinated rapidly with those of the opposite mating type (called 21-cells) even without exposure to pheromone because the sexual agglutinins are synthesized and expressed constitutively in this species. It was possible to isolate from the 5-cell surface a large glycoprotein, called 5-factor, which was released by the protease, subtilisin, and which agglutinated 21-cells (Figure 10.3A). The properties of this proteoglycan are summarized in Table 10.1. It has a molecular weight of about 1 million and contains 85% (by weight) mannan, 10% protein, and about 5% esterified phosphate. Approximately 55% of the amino acyl residues are seryl residues and 5% are threonyl residues. About half of these residues are glycosylated. The phosphate was shown to be of no biological significance, since mutants that lacked the ability to introduce phosphate into the cell wall and into the agglutinin exhibited full biological activity. The activity was destroyed by the proteolytic enzyme, pronase, and also by an exomannanase that removes terminal mannosyl residues from the mannan glycoprotein complex. Sodium periodate, which selectively destroys carbohydrate, also destroyed the agglutinating activity. These results lead to the conclusion that the integrity of both the protein and the carbohydrate are required for biological activity. Since the agglutination activity was not lost upon exposure to heat (100°C) or to acid (0.1 N HCl) it was concluded that the recognition site of the molecule might be carried by the carbohydrate moieties.

When this multivalent glycoprotein was exposed to a disulfide reducing agent such as mercaptoethanol or dithiothreitol, agglutinability was lost. Small gly-

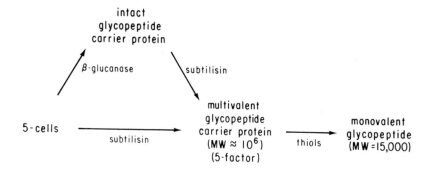

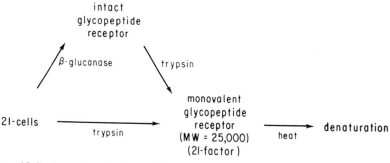

Figure 10.3. Procedures for the isolation of active components of the sexual agglutination factors present on the surfaces of haploid *Hansenula wingei* cells of opposite mating type.

Table 10.1. Properties of Protease-Released Sexual Agglutinins from *Hansenula wingei*

	5-factor	21-factor
Molecular Weight	10^6	25,000
% Carbohydrate	85% (mannan)	< 5%
% Protein	10%	>95%
	(55% Serine; 5% Threonine)	(glu and asp high)
% Diesterified phosphate	5%	< 1%
Inactivated by Pronase	+	+
Inactivated by exomannanase or periodate (0.3 mM)	+	−
Inactivated by S-S reduction	+	−
Inactivated by heat (100°C) or acid (0.1 N HCl)	−	+
Valency	Multivalent	Monovalent

copeptides were released, consisting of about 28 amino acyl residues and about 60 carbohydrate units. These fragments of uniform size bound to the surface of 21-cells and neutralized the isolated 21-factor (see below) but did not promote 21-cell agglutination. It must therefore be concluded that disulfide reduction released agglutination-specific binding fragments that carry the recognition site to which the receptor, associated with the 21-cell surface, binds. The presumed structure of 5-factor is depicted in Figure 10.4.

Treatment of 21-cells with trypsin released a monovalent 5-factor binding protein containing little carbohydrate (Figure 10.3B). This protein bound to the surfaces of 5 cells, neutralized the agglutinating activity of isolated 5-factor toward 21-cells, and was presumed to be the 5-factor recognition protein on the surface of 21-cells. It exhibited the properties summarized in Table 10.1. The protein was small and acidic and contained little carbohydrate, but a large percentage of glutamate and aspartate residues were present in the polypeptide backbone. The biological activity of the molecule was destroyed by pronase, but not by exomannanase or periodate. It was inactivated by heat and acid but not by disulfide reducing agents. Thus, only the protein moiety of this molecule is probably of biological significance, and it presumably functions as the receptor—a monovalent carbohydrate binding protein with high affinity for the carbohydrate moieties of the small receptor-specific mannopeptide released from 5-factor during disulfide reduction. A schematic depiction of 21-factor is shown in Figure 10.4.

Agglutinating factors could be released from the surfaces of a variety of yeast cells by digestion of the cell wall glucan with the enzyme, β-glucanase. The

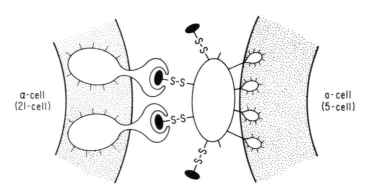

α-cell
(21-cell)

a-cell
(5-cell)

Figure 10.4. Proposed structures and modes of action of the sexual agglutination factors in yeast. The mechanism depicts complementary macromolecules on the surfaces of the two haploid cells of opposite mating type. The monovalent mannopeptide receptor on the α cell surface binds a monovalent glycopeptide fragment, many of which are bound via disulfide linkage to the mannopeptide carrier associated with the **a** cell surface. The receptor is thought to be enmeshed in the glucan layer of the α cell wall, while the mannopeptide carrier is probably associated in a similar fashion with the glucan layer of the **a** cell wall.

intact glycoprotein agglutinins isolated by this procedure resembled those isolated by limited proteolysis except that they were larger, and 21-factor contained much more carbohydrate in the form of mannan. Thus, it can be tentatively concluded that the carbohydrate moiety of 21-factor functions primarily as an anchor, holding the active receptor moiety to the surface of 21-cells (Figure 10.4).

Comparative studies of the agglutinating factors released from yeast cells have revealed that 5-factor from *H. wingei* corresponds to the sexual agglutinin present on the surfaces of α-factor-induced **a** cells of *Saccharomyces cerevisiae*, while 21-factor from *H. wingei* is structurally related to the complementary protein receptor found on the surface of **a**-factor-induced α cells of *S. cerevisiae*. Nevertheless, the homologous agglutinating activities in various yeast species show little or no cross-species reactivities or immunological relatedness as would be expected for a species-specific sexual agglutination process. It seems that while the mechanism of sexual agglutination appears to be very similar in these two yeast species, the macromolecules involved are sufficiently different to prevent cross species recognition. This postulated recognition mechanism, involving a mannopeptide-binding receptor on the surface of the α-cell, and a target mannopeptide, covalently linked via disulfide bonds to the *agglutinin carrier* on the surface of the **a** cell, is shown in Figure 10.4.

Developmental Control of Cellular Cohesion in *Dictyostelium discoideum*

As discussed in Chapter 4, the cellular slime molds, such as *Dictyostelium discoideum*, respond to starvation conditions by forming spore-filled fruiting bodies. One of the first steps initiating the transition of amoebae from the unicellular, noncohesive, vegetative state to the cohesive multicellular state involves the synthesis of two developmentally regulated carbohydrate-binding proteins. These molecules, called discoidin I and discoidin II, are thought to mediate the intercellular recognition/adhesion process that, in part, allows amoebae to aggregate to the multicellular state.

Carbohydrate-binding proteins have been isolated from a variety of related slime mold species, and their carbohydrate-binding specificities, as well as their physicochemical properties, have been examined. The properties of discoidins I and II are summarized in Table 10.2. The two lectins are tetrameric proteins of 100,000 molecular weight with different amino acid compositions. There is no carbohydrate associated with these proteins, but they bind to carbohydrate chains of glycoproteins on the surfaces of aggregation-competent slime mold cells. As shown in Table 10.2, both discoidins I and II exhibit specificity for terminal galactosyl residues, but the simple monosaccharides and galactosides that effectively inhibit binding of the lectins to the cell surface are different. Of the sugars tested, N-acetyl galactosamine most effectively binds to discoidin I while lactose most strongly interacts with discoidin II. This fact reflects structural

Table 10.2. Properties of Discoidins I and II, Carbohydrate Binding Proteins from *Dictyostelium discoideum*

Lectin	MW	# Subunits (valency)	Isoelectric point	Amino acid composition	Sugar specificity
discoidin I	100,000	4	6.1	asp>thr>glu>ser 36 24 20 15	galNAc>Meβ-gal>lac>Meα-gal
discoidin II	100,000	4	6.5	asp>ser>glu>thr 36 23 13 12	lac>Meβ-gal>galNAc>Meα-gal

features of the different carbohydrate-binding sites on the surfaces of the two lectin molecules.

The times of appearance of discoidin I and discoidin II during the early stage of fruiting body formation are depicted in Figure 10.5. Synthesis of the two proteins first occurs a couple of hours after exhaustion of the food source and continues at an exponential rate. After about 8 hr, however, discoidin II synthesis ceases while the rate of discoidin I synthesis is greatly enhanced. Thus, aggregation-competent cells possess 10- to 20-fold as much discoidin I as discoidin II. Enhanced synthesis of these two lectins has been shown to be due to enhanced rates of mRNA synthesis.

Lectins have been isolated from six different species of slime molds. Each species has two lectins, a major and a minor one, and the major lectin may be heterogeneous, possessing subclasses. All have been found to possess subunit molecular weights between 20 and 30,000. All of them exhibit specificity for carbohydrate chains with a sugar of the galactosyl configuration in the nonreducing terminal position. However, each exhibits different oligosaccharide specificities, and each agglutinates the homologous cell type at least tenfold more efficiently than any of the heterologous cell types.

Several lines of evidence suggest that these species-specific lectins function as cell surface recognition molecules during slime mold aggregation in the early stages of fruiting. First, by using antibodies directed against the pure lectins, it could be shown that they are present on the cell surface, and their appearance coincides with the acquisition of aggregation competence. Second, the sugar specificities for inhibition of lectin binding to the surfaces of slime molds are the same as those for inhibition of lectin induced hemagglutination. Third, the purified lectins enhance aggregation of cohesive, but not vegetative slime mold amoebae. Fourth, the binding affinity of the lectin for the homologous slime mold species is about tenfold greater than for a heterologous species (i.e., Kd $= 10^{-9}$ M for the homologous cell compared with 10^{-8} M for a heterologous

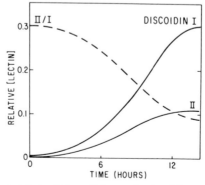

Figure 10.5. Developmental regulation of the synthesis of the two carbohydrate-binding proteins, discoidins I and II. Relative concentrations of the two lectins are plotted versus time during development after initiation of starvation conditions.

one). Finally, a mutant of *D. discoideum* that cannot differentiate past the loose aggregate stage was found to have a defect in discoidin I. These results taken together provide substantial evidence for the involvement of lectins in the aggregation process.

Further studies have revealed that not only the lectins, but also high-affinity oligosaccharide-containing cell surface receptors, to which the lectins may bind to promote cell aggregation, appear during development. One candidate for this receptor is an integral membrane glycoprotein that has been purified from *Dictyostelium*. Although discoidin I binds to this glycoprotein with high affinity, it is not clear that it functions as the physiologically relevant receptor.

The vast majority of discoidin I associated with the aggregation-competent slime mold cell is present intracellularly. However, exposure of cells to polyvalent glycoproteins such as the glycoprotein receptor elicits the appearance of additional lectin on the surface. It is possible that cross binding of the tetravalent lectin with another aggregation-competent cell is the normal signal for secretion of enhanced amounts of the lectin from a cytoplasmic store to the cell surface, and that this secretion event is necessary for high-affinity intercellular cohesion.

The presumed mechanism of lectin-mediated cell aggregation in *Dictyostelium* is illustrated in Figure 10.6. The tetravalent lectin bridges the gap between oligosaccharide-containing receptors on the surfaces of aggregation-competent cells. As may be true for any homotypic adhesive process, the arrangement of macromolecules on the surfaces of the two homologous cell types is symmetrical.

The results summarized in this section provide strong evidence for an involvement of lectins in slime mold aggregation. This evidence is not conclusive, however, particularly because the vast majority of discoidins I and II are found in the cytoplasm rather than on the cell surface. A second possible function of these molecules may involve intracellular recognition. For example, the carbohydrate moieties of glycoproteins might function as chemical "tags", which

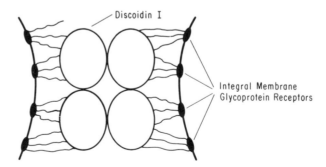

Figure 10.6. Presumed arrangements of the oligosaccharide receptor and tetravalent discoidin I on the surfaces of aggregation-competent *Dictyostelium discoideum* cells after intercellular contacts have been established.

in conjunction with the lectins direct newly synthesized glycoproteins to specific cellular organelles. More information will be required to define the physiological functions of these molecules.

Heterotypic Adhesion of Bacterial Viruses to Their Hosts and Phage-Mediated Receptor Conversion

There is extensive evidence that viruses attach to their host cells by means of receptor-mediated heterotypic recognition processes involving complementary macromolecules. The first step in viral infection is generally the adsorption of the virus to the host cell surface. Adsorption of a bacterial virus (a bacteriophage) to its host cell depends on the presence of attachment sites on the surface of the bacterium and a complementary site on the phage, which serves as the recognition unit. Evidence for an involvement of complementary macromolecules in phage adsorption initially came from studies showing that phage could adhere to heat-killed bacteria, and X-ray-inactivated phage could still adsorb to bacteria. It is now known that phage receptors on the bacterial surface are usually specific proteins, polysaccharides, lipopolysaccharides, or components of the peptido-glycan layer of the cell wall. Bacterial mutations that render a host resistant to phage attack frequently alter the cell surface macromolecule that functions as the receptor. Such mutations presumably destroy complementarity with the phage organelle of adhesion. Analogous phage mutants that adsorb to an altered cell surface receptor have been isolated. In these mutants, the phage organelle of attachment is presumably altered.

Phage with contractile tail sheaths (*E. coli* phages T2 and T4) initially attach to their bacterial hosts by means of long tail fibers that recognize specific at-tachment sites in the lipopolysaccharide layer of the Gram-negative cell. Isolated lipopolysaccharide fragments retain the capacity to bind to the phage tail fibers, and specific sugar residues in the lipopolysaccharide, which must be of the proper anomeric configuration, are crucial to this process. Moreover, since binding occurs rapidly at $0°C$, enzyme catalysis is probably not involved. The protein in the tail fiber that recognizes a specific sequence of sugar residues in the bacterial lipopolysaccharide must be a complementary sugar-binding protein.

The lipopolysaccharide (O-antigen) receptors for phages that adsorb to certain *Salmonella* hosts have been described in detail. For example, phages ε^{15} and ε^{34} require the O-antigen repeating sequence: mannosylrhamnosylgalactose. For phage ε^{15} the galactosylmannose linkage must be of the α-anomeric configuration while that for phage ε^{34} must be of the β-anomeric configuration.

During infection, O-antigen specific phages typically hydrolyze bonds within the O-antigen chain, thereby destroying the receptor. In all cases the enzymatic activity is a property of the tail-like attachment protein. That is to say, adsorption of this type of phage may involve the formation of an enzyme–substrate complex in which a certain integral part of the phage tail is the enzyme and the O-antigen receptor is the substrate. Even in the absence of the virion, isolated attachment proteins can adsorb to the O-antigen moiety of the lipopolysaccharide and carry out enzymatic cleavage. For example, the tail of bacteriophage P22 has an endorhamnosidase activity that hydrolyzes rhamnosylgalactose linkages in the *S. typhimurium* O-antigen chain. Bacteriophage ε^{15} has a similar endorhamnosidase, which is specific for the O-antigen of *S. anatum*. Certain mutants of phages P22 and ε^{15}, having temperature sensitive genetic defects, exhibit decreased endorhamnosidase activity and are noninfectious at the nonpermissive temperature even though they adsorb normally to their host cells. Evidently, the attachment proteins of such mutants have retained their O-antigen binding specificity while losing their endorhamnosidase catalytic activity.

Bacteriophage P22 adsorbs to its host by a process that is typical of *Salmonella* O-antigen-specific phages and is thought to be multistepped. The tail of phage P22 consists of six tail parts, each of which has a molecular weight of about 170,000 and contains two identical polypeptide chains. The isolated tail parts interact reversibly with *S. typhimurium* cells and cleave the O-antigen chains. Intact phage, however, bind irreversibly to *S. typhimurium,* and initial binding is linearly dependent on the number of tail parts per phage. The presence of only one tail per head is sufficient to allow attachment of the phage to its host, and such tail-deficient phage can hydrolyze O-antigen. They are not infectious, however, because they cannot eject their DNA. Furthermore, binding to the host cell is reversible. In contrast to adsorption, DNA ejection was found to be proportional to the third power of the number of tail parts per head. This fact indicates that three tail parts per head participate in DNA ejection. From these observations it appears that a multistep model for P22 infection operates as follows: Initial contact between the phage and the bacterial cell probably occurs between one of the tail parts and an O-antigen chain. Following cleavage of a rhamnosylgalactose bond, the phage would either (1) continue down the chain to which it was attached, (2) dissociate and become free, or (3) dissociate and bind nearby. Repeated cleavage of and binding to the O-antigen chain may allow the phage to approach the cell surface, at which point a second interaction involving at least three tail fibers might occur. Most likely this second interaction elicits a third reaction involving a slender fiber connected to the neck of the virus and a target on the cell surface responsible for initiating DNA ejection. In the case where only one or two tail parts are present per head, the phage might be unable to orient itself properly for the fiber–target interaction.

Studies of bacteriophage ε^{15} infectivity also support a multistep adsorption process. Adsorption of ε^{15} phage to *S. anatum* cells is relatively insensitive to temperature and pH although O-antigen hydrolysis is temperature dependent and optimal at a slightly acidic pH. DNA ejection occurs only within a narrow pH range near neutrality, is very temperature dependent, and requires the presence of divalent cations. Two classes of temperature sensitive mutants of ε^{15} have been isolated which, at the nonpermissive temperature, either (1) adsorb normally to cells but do not degrade O-antigen or eject DNA, or (2) absorb reversibly, degrade large amounts of O-antigen, but do not eject their DNA. The evidence suggests that adsorption is essential for O-antigen degradation and that O-antigen degradation is essential but not sufficient for DNA ejection.

In the case of temperate bacteriophage such as ϵ^{15} and P22, the infected host cell may harbor the prophage for many generations, and the prophage genetic material may direct some of the biosynthetic activities of the cell. Cell surface characteristics may change, and this phenomenon is referred as *conversion*. Several *Salmonella* phage, for example, carry genes that encode proteins responsible for alterations in the O-antigen side chains of the bacterial cell surface lipopolysaccharide. Among the best characterized of these viruses is phage ϵ^{15}. The normal host of phage ϵ^{15} possesses O-antigen chains with galactosyl residues that are of the α-anomeric configuration and are O-acetylated. Upon conversion, the acetyl groups are lost, and the α-anomeric linkage is converted to β. Three phage genes are involved. One gene probably encodes a transcriptional repressor protein that prevents synthesis of the acetylating enzyme. A second phage gene encodes a small, heat stable protein that apparently inactivates the host cell α-polymerase; and the third phage gene involved in conversion encodes a β-polymerase that polymerizes the repeating trisaccharide with the β-anomeric configuration. In this way the phage alters its own receptor, preventing superinfection by homologous phage. Phage conversion can therefore be considered to be functionally analogous to processes that prevent fusion of more than two haploid yeast cells of opposite mating type, or of more than one sperm cell to the haploid egg cell.

The genetic and biochemical tools available for the dissection of phage adsorption and conversion render this system one of the best available for detailed analyses. Future studies may reveal the structural and functional requirements of the cell surface macromolecules mediating adhesion as well as the sequence of receptor-mediated events that eventually lead to a biological response—in this case the injection of phage nucleic acid into the host cell. It would not be surprising if the receptor is multifunctional in this regard.

Cell–Cell Contact as a Signal for Induction of Tissue-Specific Protein Biosynthesis

During embryogenesis new cell types develop with unique but predictable programs of gene expression. These programs are necessitated by the fact that different cell types perform distinct physiological functions and therefore must possess unique sets of tissue-specific proteins. The mechanisms of induction of these proteins during embryological development are generally unknown, but numerous transplantation studies have shown that cell–cell interactions influence expression of the program.

One such tissue-specific protein is a Ca^{2+} binding protein known as the S100 protein. It is a strongly acidic protein of about 25,000 molecular weight, rich in aspartyl and glutamyl residues and consisting of three nonidentical subunits. It is localized largely to nervous tissue, having a concentration in brain more than 1000-fold that found in any other body tissue examined. It is probably restricted to the glial elements in the brain, and may play an important role in brain function since its level increases dramatically during functional maturation of the nervous system. One study has suggested that it mediates learning since its concentration increases during learning, and injection of anti-S100 protein antibody interferes with the learning process.

The synthesis of the S100 protein has been studied in cloned populations of a rat glioma cell line, a tumorigenic line derived from glial cells in the brain. This transformed cell line synthesizes the S100 protein in response to highly specific stimuli. The protein is not made in appreciable amounts when the cells are grown in sparse cultures. However, when the growing glial cells come into direct contact, rates of S100 protein synthesis are greatly enhanced. This occurs several generations before cell growth ceases, during a time when net protein synthesis is beginning to slow. The enhanced rate of S100 protein synthesis is due to the accumulation of messenger RNA coding for the protein. No such increase is observed when the glioma cells come into contact with other cell types such as fibroblasts. This observation suggests that highly specific homotypic cell–cell interactions promote transcription of the genes encoding the S100 protein. Some evidence for the involvement of a cell surface receptor has been obtained, and inhibitors of intracellular microtubule formation block the inductive effect. Thus, transmission of the signal from the cell surface to the nucleus might involve microtubules.

Contact Inhibition of Fibroblast Growth

When normal animal cells maintained in tissue culture come into contact, two biological responses are usually observed. First, there is a local and rapid inhibition of the elaborate ruffling movements associated with the advancing edge of the cell membrane, and subsequently cell motility ceases. This phenomenon has been termed *contact inhibition of motility*. Second, when the cell density becomes higher so that the culture is confluent, the cells cease to divide and arrest in the G1 phase of the cell cycle. This phenomenon has been termed *contact inhibition of growth*. In other cases, heterotypic intercellular associations have been shown to provide a positive signal for the stimulation of cell growth. Thus, neurites stimulate the proliferation of Schwann cells. Because cancer cells sometimes lose sensitivity to contact inhibition, the phenomenon has been studied in some detail. In this section the available information will be summarized as an example of a complex response to homotypic adhesion.

The phenomenon of density-dependent inhibition of fibroblast growth was first recognized about 20 years ago as a result of observations made with cells growing in the tissue culture environment. Contact inhibition occurs even if the growth medium is frequently changed, showing that growth stasis does not result from

the depletion of medium components. Although growth factors do influence the cell density at which growth arrest occurs, decreased access of the cells to growth-stimulatory agents as a result of cell crowding cannot, by itself, account for the phenomenon.

A variety of factors and conditions influence the decision to arrest in G1 or to pass through this phase of the cell cycle and initiate a round of DNA replication. Thus, nutrient availability, mitogenic hormones, and growth factors promote cell division, while tissue-specific growth-inhibitory agents called chalones and cell–cell contacts exert negative control over growth. Possibly there is an intracellular mechanism for integrating the input signals from all of these agents facilitating the decision-making process.

In order to test the possibility that receptors in the plasma membrane mediate contact inhibition of growth, mouse fibroblast plasma membrane vesicles were isolated and incubated with sparse, actively growing cultures of mouse fibroblasts. A large percentage of the cells became growth arrested in the G1 phase of the cell cycle. Growth arrest was fully reversible upon removal of the membrane vesicles. Vesicles prepared from confluent fibroblasts were more effective in inhibiting cell growth than those isolated from actively growing cells. As expected, the inhibiting vesicles appeared to be derived from the plasma membrane. Consistent with these observations, vesicles derived from plasma membranes of transformed fibroblasts, which do not exhibit density-dependent growth inhibition, were not effective in inhibiting actively growing cultures of normal fibroblasts.

When vesicles were solubilized with the detergent, octylglucoside, a biologically active growth inhibitory factor was found to have survived the treatment. Like the plasma membrane vesicles, the extracts reversibly inhibited DNA synthesis, and this inhibitory effect was reversed upon addition of growth factors. It was found that inhibition of growth in the presence of a mitogenic protein did not prevent binding of the mitogen to the cell surface; it merely antagonized its effect on cell growth. The inhibitory factor was found to lose its activity during incubation at elevated temperature and was presumed to be a protein. A distinct membrane protein was found to inhibit the activity of an amino acid transport system, the A system, which has been extensively implicated in growth regulation of animal cells. Possibly this transport system represents one component of the contact inhibition signal, and its activity may serve as one of several biochemical systems that integrate mitogenic and inhibitory signals influencing the decision to arrest at or continue through the G1 phase of growth.

Sea Urchin Egg Fertilization

One of the most complex programs of biochemical events triggered by heterotypic cell–cell contacts involves activation of a dormant developmental program in the egg cell upon fertilization by a sperm cell (Figure 10.7). The most detailed information concerning fertilization has come from studies of sea urchins, pri-

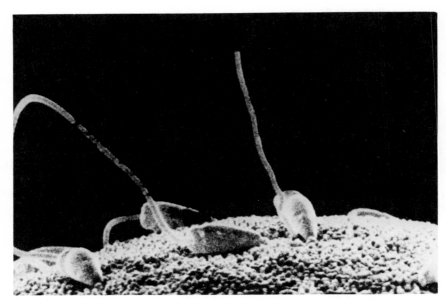

Figure 10.7. Electron micrograph of sea urchin sperm cells adhering to the surface of the much larger egg cell. From D. Epel, "The Program of Fertilization," Copyright © 1977 by Scientific American, Inc. All rights reserved. Courtesy of Dr. William Byrd, University of Texas Medical School, Houston, Texas.

marily for technical reasons. Female sea urchins produce about 500 million eggs per season while males generate about 100 billion sperm cells. Fertilization is analogous to mating in yeast and is similarly initiated by heterotypic association due to the presence of complementary macromolecules on the surfaces of these two germ cells.

The sea urchin egg plasma membrane is surrounded by a vitelline envelope with an extracellular layer of jelly outside the vitelline. Before sea urchin gametes can fuse, the sperm must swim through the jelly layer, undergo a series of events known as the acrosome reaction, and bind tightly to the vitelline envelope. Following plasma membrane fusion, the egg membrane undergoes a series of changes in ionic permeability and electrical potential that strongly inhibit fusion of additional sperm cells with the egg plasma membrane. An exocytotic event called the cortical reaction, which occurs more slowly, gives rise to structural changes that further inhibit sperm entry. These biochemical changes have become known as the *block to polyspermy*. As the single internalized sperm nucleus moves to the center of the egg to fuse with the egg pronucleus, the egg abandons its quiescent state and becomes metabolically active.

The jelly coat of the sea urchin egg has been isolated and characterized in order to study its effects on sperm. The major components of jelly are a complex fucose sulfate polymer (about 80%) and a sialoprotein (about 20%). The jelly also contains tiny amounts of a small peptide called speract, an extremely potent activator of sperm adenylate cyclase (the cyclic AMP biosynthetic enzyme) and of sperm respiration and motility. Speract, however, does not induce the sperm acrosome reaction, and this reaction apparently depends on the fucose sulfate portion of jelly or some other component of jelly, as yet undiscovered.

The acrosomal granule of the sperm represents an organelle highly specialized for interaction with the egg. A Golgi-derived membrane-bound structure, it is believed to contain lytic enzymes for dissolving egg extracellular material, a species-specific glue-like protein, called bindin, unpolymerized actin, and possibly new membrane and/or membrane proteins. The acrosome reaction involves fusion of the acrosomal and plasma membranes and the resultant release of the acrosome granular contents into the external medium. Subsequently, a long, filamentous structure called the acrosome process, consisting of polymerized actin, is formed. The new membrane material covering the acrosomal filament is highly fusigenic and will be the part of the sperm that fuses with the egg.

Numerous biochemical changes occur in the sperm during the acrosome reaction. Early events include uptake of sodium coupled to hydrogen ion efflux, raising the internal pH of the sperm, as well as an influx of calcium ions. These two events can be separated by manipulating external ion concentrations and by using various ionophores and channel blocking compounds. Evidence of this type suggests that alkalinization of the internal pH triggers actin polymerization while fusion of the acrosomal vesicle and plasma membrane requires increased intracellular calcium. Cyclic AMP levels also rise during the acrosome reaction, and cyclic AMP might function as a second messenger to activate a protein kinase in the sperm cytoplasm. Shortly thereafter an ion transport system is activated, which catalyzes the massive efflux of K^+ from the cell. It is possible that KCl exits from the cell via a furosemide-sensitive neutral salt transport system that is activated by a cyclic AMP-activated phosphorylation mechanism. Loss of cytoplasmic KCl may cause cell shrinkage and facilitate extrusion of the sperm nucleus into the egg. The specific sperm cell surface components involved in induction and propagation of the acrosome reaction have not yet been isolated, but preliminary evidence suggests that a glycoprotein of 84,000 molecular weight may act as the jelly receptor.

Species-specific binding of sea urchin sperm to eggs is mediated by bindin, a major constituent of the acrosomal granule. Immunological studies have shown that bindin coats the extended acrosomal filament and is located between the filament and the egg vitelline envelope. Purified bindin, a nonglycosylated protein of molecular weight 30,000, will aggluntinate unfertilized (but not fertilized) eggs in a species-specific manner. This agglutination process can be inhibited by a glycopeptide, released from eggs by protease digestion, which may be part of the bindin receptor molecule. The glycopeptide is also liberated by natural proteases extruded from the egg at fertilization. Bindin-induced egg agglutination can be inhibited by several artificial sugar complexes that contain fucose and xylose, suggesting that a lectin-like interaction may be responsible for species-specific sperm–egg binding. A mechanistic analogy to sexual agglutination of haploid yeast cells seems likely.

As noted above, fusion of the egg and sperm plasma membranes is followed within a second or two by a rapid influx of sodium ions (Figure 10.8), and this unidirectional flux of positive charge depolarizes the egg cell, thereby changing the cell surface properties so that further sperm penetration is inhibited. This process is called the fast block to polyspermy. After about 20 seconds, the permeability of the membrane to Ca^{2+} increases dramatically, and the intracellular concentration of Ca^{2+} transiently increases and then decreases many-fold, all within about a minute. Another transport system that catalyzes Na^{+}/H^{+} exchange appears to be activated by Ca^{2+}, and a massive uptake of Na^{+}, coupled to H^{+} extrusion, ensues (Figure 10.8). H^{+} extrusion gives rise to a pH gradient across the membrane with the internal pH more basic than the pH of the medium. Shortly thereafter various metabolic enzymes are activated, oxygen consumption increases, and a reducing environment is established. Five minutes after the time of fertilization, transport systems are activated, bringing amino acids, phosphate, nucleosides, and other nutrients into the cell, and the protein biosynthetic machinery of the cell begins to operate. After an additional 15 min, DNA synthesis is initiated, and cell division ensues after about 90 min.

Some of these events appear to be causally related. Thus, the initial influx of Na^{+}, which depolarizes the membrane, probably causes Ca^{2+} channels in the membrane to open transiently. Experiments with large fish eggs have shown that Ca^{2+} entry starts at the point of sperm entry; a wave of Ca^{2+} permeability is propagated over the entire cell surface, and then this Ca^{2+} flux is extinguished. The unidirectional wave of Ca^{2+} permeability is attributed to the sequential opening and closing of Ca^{2+} channels within the membranes of the cell. A similar excitable process may be responsible for Ca^{2+} entry in the sea urchin egg. Part

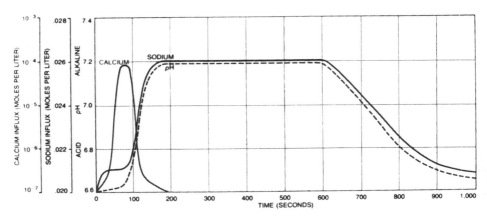

Figure 10.8. Time course for the changes in the cytoplasmic ionic composition of the sea urchin egg following fertilization. The figure depicts changes in Na^{+}, Ca^{2+} and H^{+} concentrations. From D. Epel, "The Program of Fertilization," Copyright © 1977 by Scientific American, Inc. All rights reserved.

of the Ca^{2+} released into the cytoplasms of these cells may be derived from intracellular stores.

The transient increase in free cytoplasmic Ca^{2+} apparently activates the Na^{+}/H^{+} exchange transport system, the activity of which renders the internal pH more basic. It is probably the alkaline cytoplasmic pH that activates biosynthetic and transport activities that eventually lead to cell division and embryogenesis. This scheme of causally related events is depicted in Figure 10.9. In addition, there are many molecular events not shown in Figure 10.9 that are triggered by fertilization and represent essential components of the embryogenic process. It is worth reemphasizing, however, that cell surface heterotypic reception (the egg–sperm interaction) is the trigger for activation of a series of transport systems that change the ionic composition of the cytoplasm, the internal pH, and the membrane potential. Thus, egg–sperm adhesion serves as the reception process; cytoplasmic ions and the transmembrane electrical potential (and possibly cyclic AMP in sperm) function as second messengers; and the ultimate biological targets of second messenger action are metabolic and biosynthetic enzymes. This relay system must be one of the most complex in nature. These considerations show

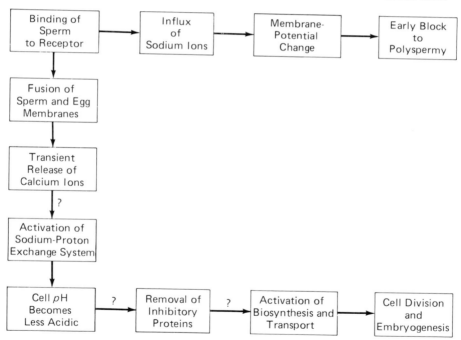

Figure 10.9. Parallel sequences of events, initiated by sperm-egg reception, which lead to the early block to polyspermy (top) and embryogenesis (bottom). The causal relationships implied in the figure have not been established in all cases, and other related events not shown are known to occur concomitantly. Thus, the scheme shown is simplified and tentative.

that regulatory processes responsible for the most intricate biological programs, such as the initiation of embryogenesis in multicellular organisms, can be understood in terms of the same processes that have long been recognized in much simpler microorganisms.

Selected References

Ballou, C.E. "Yeast Cell wall and Cell surface" in *The Molecular Biology of the Yeast Saccharomyces, Vol. II: Metabolism and Gene Expression* (J.N. Strathern, E.W. Jones, and J.R. Broach, eds.), Cold Spring Harbor Laboratories, Cold Spring Harbor, New York, 1982.

Bartles, J.R., B.C. Santoro, and W.A. Frazier. Discoidin I-membrane interactions III. Interactions of Discoidin I with living *Dictyostelium discoideum* cells, *Biochim. Biophys. Acta 687:*137–146 (1982).

Bennett, D., The T-locus of the Mouse, *Cell 6:*441–454 (1975).

Bunge, R., L. Glaser, M. Lieberman, D. Raben, J. Salzer, B. Whittenberger, and T. Woolsey. Growth control by cell to cell contact, *J. Supramolec. Struc. 11:*175 (1979).

Edelman, G.M. and C-M. Chuong. Embryonic to adult conversion of neural cell adhesion molecules in normal and staggerer mice, *Proc. Natl. Acad. Sci. USA 79:*7036–7040 (1982).

Epel, D. Mechanisms of activation of sperm and egg during fertilization of sea urchin gametes, *Curr. Top. Dev. Bio. 12:*185 (1978).

Epel, D. "Experimental Analysis of the Role of Intracellular Calcium in the Activation of the Sea Urchin Egg at Fertilization" in *The Cell Surface: Mediator of Developmental Processes* (S. Subtelny and N.K. Wessels, eds.), Academic Press, 1980.

Gluecksohn-Waelsch, S. Genetic control of morphogenetic and biochemical differentiation: Lethal albino deletions in the mouse, *Cell 16:*225 (1979).

Graham, C.F. "Teratocarcinoma Cells and Normal Mouse Embryogenesis" In *Concepts in Mammalian Embryogenesis* (M.I. Sherman, ed.), The MIT Press, Cambridge, Massachusetts, 1977.

Hausman, R.E. and A.A. Moscona. Immunological detection of retina cognin on the surface of embryonic cells, *Exptl. Cell Res. 119:*191–204 (1979).

Hausman, R.E. and S.G. Velleman. Prostaglandin E-1 receptors on chick embryo myoblasts, *Biochem. Biophys. Res. Comm. 103:*213–218 (1981).

Hoffman, S., B.C. Sorkin, P.C. White, R. Brackenbury, R. Mailhammer, U. Rutishauser, B.A. Cunningham, and G.M. Edelman. Chemical characterization of a neural cell adhesion molecule purified from embryonic brain membranes, *J. Biol. Chem. 257:*7720–7729 (1982).

Jacob, F. Mouse teratocarcinoma and mouse embryo. *Proc. R. Soc. Lond. B. 201:*249–270 (1978).

Jones, G.W. and R.E. Isaacson. Bacterial adhesions and their receptors, *CRC Crit. Rev. Micro. 10:*229–260. (1983).

Labourdette, G., J.B. Mahony, I.R. Brown, and A. Marks. Regulation of synthesis of a brain-specific protein in monolayer cultures of clonal rat glial cells, *Eur. J. Biochem. 81:*591 (1977).

Lieberman, M.A. and L. Glaser. Density-dependent regulation of cell growth: An example of a cell–cell recognition phenomenon, *J. Membrane Biol. 63:*1 (1981).

Loomis, W.F., ed. *Development of Dictyostelium*, Academic Press, New York, 1980.

Magnuson, T. and C. J. Epstein. Genetic control of very early mammalian development, *Biol. Rev. 56:*369 (1981).

Martin, G.R. Teratocarcinomas and mammalian embryogenesis, *Science 209:*249–270 (1980).

McKinnell, R.G., M.A. DiBerardino, M. Blumenfeld, and R.D. Bergad, eds. *Results and Problems in Cell Differentiation. Vol. II. Differentiation and Neoplasia*, Springer-Verlag, New York, 1980.

Moscona, A.A. "Surface Specification of Embryonic Cells: Lectin Receptors, Cell Recognition and Specific Cell Ligands" In *The Cell Surface in Development* (A.A. Moscona, ed.), John Wiley and Sons, Inc., 1974.

Olden, K., J.B. Parent, and S.L. White. Carbohydrate moieties of glycoproteins. A reevaluation of their function. *Biochim. Biophys. Acta 650:*209 (1982).

Paigen, K. Temporal Genes and Other Developmental Regulators in Mammals in *The Molecular Genetics of Development* (T. Leighton and W.F. Loomis, eds.), Academic Press, New York, 1980.

Randall, L.L. and L. Philipon, eds. *Virus Receptors (Receptors and Recognition, Series B*, Vol. 7), Chapman and Hall, London 1980.

Reitherman, R.W., S.D. Rosen, W.A. Frazier, and S.H. Barondes, Cell surface species-specific high affinity receptors for discoidin: Developmental regulation in *Dictyostelium discoideum, Proc. Natl. Acad. Sci. USA 72:*3541–3545 (1975).

Rosen, S.D., J.A. Kafka, D.L. Simpson, and S.H. Barondes. Developmentally regulated, carbohydrate-binding protein in *Dictyostelium discoideum, Proc. Natl. Acad. Sci. USA 70:*2554 (1973).

Shapiro, B.M., Schackman, R.W. and C.A. Gabel, Molecular approaches to the study of fertilization. *Ann. Rev. Biochem. 50:*815–843 (1981).

Sherman, M.I. and D. Solter, eds. *Teratomas and Differentiation*, Academic Press, New York, 1975.

Thorner, J. "Intercellular Interactions of the Yeast *Saccharomyces cerevisiae*" in *Microbial Differentiation* (T. Leighton and W.F. Loomis, eds.), Academic Press, New York, 1980.

Wessells, N.K. *Tissue Interactions in Development*, An Addison-Wesley Module in Biology, No. 9, Addison-Wesley Publishing Co., Inc., Philippines, 1973.

CHAPTER 11

Switch Mechanisms Regulating Gene Expression

> . . . the great masters of modern science are precisely those whom nature inspires with most reverence and awe. For as their minds are wafted by their wisdom into untravelled worlds, they find new fields of knowledge expanding to the view, the firmament ever expands, the abyss deepens, the horizon recedes. The proximate Why may be discovered; the ultimate Why is unrevealed.
>
> *Winwood Reade*

In Chapter 1 we noted that differentiation-specific genes expressed during *Drosophila* development were controlled by many reversible and irreversible "switch" processes. Specific genes were sequentially activated as part of a program of differentiation. Some of these genes could be activated and silenced reversibly, depending on, for example, hormone availability, while others were activated irreversibly. Irreversible switches presumably represent decision making events or processes of determination, and they may serve as branch points or sites of regulation in the developmental program.

In virtually all living organisms, the consequences of genetic switch mechanisms have been observed, but until recently an aura of mystique has surrounded these phenomena. Switches in gene expression occur with high frequency, relative to mutation rates, and this frequency is not appreciably altered by mutagens or by gene dosage. Consequently, terms such as *epigenetics* (almost genetics) or *paramutation* (beyond mutation) have been used to describe these phenomena. They have been observed in all living organisms, from the simplest to the most complex. Thus, plants and animals, animal cells in tissue culture, eukaryotic and prokaryotic microorganisms, and bacterial viruses all show epigenetic behavior. In this chapter we shall select some of these processes for discussion and pursue them to the molecular level when possible. The examples chosen are representative of the better characterized switch processes occurring in nature.

Epigenetics in Animal Cells Maintained in Tissue Culture

Differentiation-specific genes are switched on or off at specific times during embryonic development of an organism. When grown in the tissue culture environment in pure cell populations, animal cells also exhibit epigenetic behavior, sometimes referred to as *genetic instability,* showing that the switch does not depend only on organismal or tissue integrity. Numerous nonessential, "luxury" traits have been shown to be under genetic control and heritable, but their expression can be turned on and off, or up and down, at high frequency, provided that the cell type is programmed for their expression. These switch processes are independent of mutagen concentration and cell ploidy, eliminating the possibility of a mutational event. Cellular morphology, drug resistance, pigmentation, and the synthesis of specific enzymes have all been shown to be subject to epigenetic changes in appropriate cell lines. Many genes coding for differentiated functions may be controlled independently. However there may exist *master switches* that control expression of several traits. *Intergenic repression,* where expression of one gene precludes expression of another, may also be operative. The results of some studies suggest that control may be mediated by *cytoplasmic regulatory constituents* encoded by unlinked genes. Control may be positive, involving a cytoplasmic activator of gene expression, or negative, involving a cytoplasmic repressor. In some cases threshold levels of the cytoplasmic regulatory substances may be important in the all-or-none control of gene expression. In this section, a representative example of epigenetics in tissue culture will be discussed.

Neuroblastoma cells are tumor cells derived from neurons, electrically excitable cells of the nervous system. These transformed cells frequently retain the differentiated properties of the normal cells of origin. Clones of one such neuroblastoma line were examined for a number of differentiated characteristics. These traits included (1) neurite formation and electrical excitability, (2) presence of choline acetyltransferase (CAT), the enzyme that catalyzes synthesis of acetylcholine, the neurotransmitter of cholinergic nerves, (3) presence of tyrosine hydroxylase (TH), an enzyme involved in the biosynthesis of catechols such as adrenaline and noradrenaline, the neurotransmitters of adrenergic nerves, (4) presence of acetylcholine esterase (ACE), the hydrolytic enzyme that destroys acetylcholine.

Properties of the isolated clones are summarized in Table 11.1. Among the clones were inactive ones which were not excitable and did not form neurites. These cells did not synthesize either of the enzymes involved in neurotransmitter biosynthesis. Excitable clones, on the other hand, could make either tyrosine hydroxylase or choline acetyltransferase, but not both. All cells synthesized

Table 11.1. Properties of Isolated Clones of a Neuroblastoma Cell Line

Clone type	Excitable	Neurite	CAT	TH	ACE
Inactive	−	−	−	−	+
Axon negative	+	−	−	−	+
Axon positive	+	+	−	−	+
Adrenergic	+	+	−	+	+
Cholinergic	+	+	+	−	+

acetylcholine esterase. In subsequent experiments, clones containing or lacking choline acetyltransferase were grown for 20 generations, recloned, and again assayed for the enzymes. About 20% of the clones assayed had changed their phenotype from cholinergic to noncholinergic, or vice versa.

Finally, a neurite-negative clone (which possessed low tyrosine hydroxylase and choline acetyltransferase activities) was grown for 20 generations, and clones producing large cells with neurites were isolated. While neurite-negative clones were always negative for both neurotransmitter biosynthetic enzymes, neurite-positive clones sometimes regained the ability to synthesize one of the biosynthetic enzymes. These observations led to the following conclusions: (1) Although all of the neurite-specific characteristics studied were inherited from generation to generation, expression of differentiated traits was quantitatively modulated with high frequency (about 10^{-3} per cell per generation). (2) Neurite expression appeared to serve as a master switch, controlling both choline acetyltransferase and tyrosine hydroxylase syntheses in a positive but nonobligatory fashion. This conclusion resulted from the fact that neurite-negative clones never showed appreciable activities of the biosynthetic enzymes. (3) Finally, intergenic exclusion appeared to be operative since the enzyme characteristic of only one of the neurotransmitter biosynthetic pathways was present in neurite-positive clones. These results appear to correlate with observations made in intact animals. Axons in the central nervous system may be either cholinergic or adrenergic, but not both, while nonaxonal brain cells do not synthesize neurotransmitters.

Studies of other epigenetic systems in animal cells grown in the tissue culture environment have shown that epigenetic changes are not generally due to structural gene mutations, but result from changes in the levels of the protein synthesized. Cell hybridization studies have led to the suggestion that cytoplasmic repressors and activators may control levels of these proteins by mechanisms that are highly sensitive to gene dosage. Thus, epigenetic changes in gene expression in tissue culture may be due to altered levels of transcriptional regulatory proteins.

DNA Methylation and the Control of Gene Expression in Animal Cells

Recent evidence has led to the suggestion that methylation of the DNA adjacent to certain structural genes may influence the level of expression of these genes. Moreover, a specific mechanism by which the methylation patterns of a parental cell can be passed on to its daughter cells following mitosis has been proposed. In this section the evidence for this mechanism of gene regulation will be discussed.

In animal cells the site of DNA methylation is the 5′ position of cytosine in the sequence CpG. In order to study these sequences, use has been made of two endonucleases, called *restriction endonucleases,* which cleave the DNA sequence, CCGG. One of these enzymes (termed HpaII) can cleave the unmethylated sequence, while the other enzyme (termed MspI) can cleave the sequence regardless of its state of methylation. Use of these enzymes has allowed fairly extensive determination of the methylation patterns at certain CCGG sequences.

Studies with these enzymes have shown that certain CpG sites near a regulated gene are poorly methylated in a tissue that expresses the gene, relative to one that does not express the gene. A causal relationship is likely because transformation of cells with appropriately methylated DNA does not result in normal gene expression although transformation with unmethylated DNA allows gene expression. This difference was shown not to be due to preferential uptake of the unmethylated polymer.

The methylation pattern appears to be stably inherited in a number of cell lines. Stable inheritance probably results from the fact that the methyl transferase acts efficiently on a CG site when the complementary GC site is already methylated, but it does not appreciably methylate either strand when neither site is methylated. The daughter strand of a newly replicated duplex is therefore methylated whenever the parent strand is methylated.

Inhibitors of *in vivo* DNA methylation have been identified. These compounds, including 5-azacytidine, can induce differentiation of cultured mouse embryo cells although structural analogues that do not inhibit methylation have no effect on the expression of differentiated traits.

Exactly how methylation controls gene expression is not clear. Possibly methylation alters the interactions of transcriptional regulatory proteins with the DNA, either by modifying local interactions with the active sites of sequence-specific proteins, or by introducing changes in the secondary and tertiary structure of the nucleic acid. Conformational effects of methylation on synthetic DNA strands have, in fact, been demonstrated.

While DNA methylation appears to play an important role in regulating gene expression during the development of some groups of animals, other animals, such as insects, do not exhibit similar degrees of methylation. This observation leads to the suggestion that several mechanisms for regulating of gene expression may be available to developing organisms, and the processes chosen by any two such organisms may differ either quantitatively or qualitatively.

Paramutation in Plants

Several plants have been observed to exhibit various forms of genetic instability. For example, the *rogue* trait in the garden pea gives rise to pointed leaves, upwardly curving pods and reduced stipule size. While both wild type and rogue homozygotes breed true, normal/rogue hybrids are not stable. The hybrid exhibits an intermediate phenotype during early growth (at the base of the plant) and becomes more rogue-like as development proceeds (at the tips of the plant). Unidirectional mass somatic paramutation appears to be occurring.

Among the better characterized paramutagenic systems in plants is the *R* locus in maize, which conditions anthocyanin synthesis and thus controls pigmentation in the aleurone layer of the seed endosperm. The R allele gives rise to a red aleurone layer, the r allele results in the lack of pigmentation and a colorless aleurone layer, and the R^{ST} allele promotes formation of a stippled (spotted) aleurone layer. While R is *paramutable, r* is not. R^{ST} is *paramutagenic* and can induce paramutation in the R allele.

The endosperm is triploid, deriving two chromosomal copies from the mother and one from the father. In the series *rrr, Rrr, RRr,* and *RRR,* the aleurone layer proceeds from colorless to red with the amount of pigmentation proportional to gene dosage. In the corresponding series with R^{ST} replacing the R allele, the same is true, but the aleurone layer is stippled.

Test crosses with females carrying the recessive r allele in the homozygous state (see Figure 11.1) revealed that crosses in which the R and R^{ST} alleles were not together yielded the expected phenotypes. However, when the male parent was of the $R^{ST}R$ phenotype, half of the offspring exhibited the $R^{ST}rr$ phenotype (as expected), but the other half exhibited an $R'rr$ genotype with weaker pigmentation than expected for the Rrr genotype. It appeared that the R^{ST} allele, which remained unaltered with respect to pigmentation, was paramutagenic upon the R allele. Moreover, the altered R allele, R', which gave rise to decreased pigmentation, became weakly paramutagenic.

Further studies led to the conclusion that all R alleles (including R') are inherently metastable and that the paramutability of the R allele is greatly increased by the presence of the R^{ST} allele. Surprisingly, certain mutagens appeared to counteract the effects of the R^{ST} allele, increasing pigmentation.

On the basis of these observations it was suggested that the R locus consists of two genetic regions: a structural gene complex coding for the pigment-synthesizing enzymes, and a *repressing* segment consisting of varying numbers of a common repeating unit (a metamer). The degree of repression is assumed to be proportional to the number of metamers. Directed paramutation might result from metamer replication or deletion. On the other hand, the observed variability and instability of gene expression might involve transposable elements.

rr♀ x RR♂   ⟶   All Rrr   (stable)

rr♀ x $R^{ST}R^{ST}$♂   ⟶   All $R^{ST}rr$ (stable)

                            ½ $R^{ST}rr$ (stable)
rr♀ x $R^{ST}R$♂   ⟶   +
                            ½ R'rr   (epigenetically unstable)

Figure 11.1. Test crosses revealing the paramutability of the R allele and the paramutagenic characteristics of the R^{ST} allele. In all crosses, the female (♀) provides the recessive r allele. The R and R^{ST} alleles are provided by the male (♂). R' represents a variant of R exhibiting less pigmentation than R and weakly mutagenic properties.

Transposable Elements in Plants Controlling Gene Expression and Mutation Rate

In corn, anthocyanine production and certain other traits are apparently subject to control by transposable genetic elements (transposons), which can insert into the genome at various locations and alter gene expression. Two such transposon systems have been studied in some detail. One is designated the *Ac–Ds* system, the other, the *Spm–I* system. These systems regulate gene expression and mutation rate in a fashion that may be subject to developmental control, and hence it is important to discuss them here.

Both elements can insert near a locus (designated *A*), which provides the structural gene(s) for anthocyanine pigmentation. Expression of the *A* gene is regulated by a controlling element (*Ds* for dissociable element), which is transposable within the genome and can fully or partially silence *A* when inserted within or near it. *Ds* can "change in state," and these changes influence the timing and quantity of *A*-gene silencing. A second regulatory gene (designated *Ac* for activating element) allows *Ds* to undergo transposition and change its state. In the absence of *Ac, Ds* is "locked" into a particular position and state. Like *Ds, Ac* can undergo transposition and can proliferate to yield more than a single copy per genome. Consequently, *Ac,* like *Ds,* is a transposon, and as such may be able to pair up with *Ds* so that both genetic elements are transposed as a unit.

Both the *Ac* and *Ds* transposons have been clonally isolated and sequenced. *Ac* is a 4,500 nucleotide long transposon which contains two structural genes encoding the enzymes involved in transposition of the plasmid from one position in the genome to another. As is characteristic of other transposons, the plasmid is flanked by inverted terminal repeats which are recognized and acted upon by the transposase. *Ds* possesses the same nucleotide sequence as *Ac* (including the inverted terminal repeats), but it is smaller, owing to a deletion within or extending through the transposase structural gene. Consequently, transposition of *Ds* occurs only when *Ac* is present. These features appear to be characteristic of the second, more complex, genetic regulatory system, the *Spm–I* system.

The *Spm–I* system consists of a controlling element, *I,* which affects the activities of structural genes such as *A* when inserted near or within them, as well as a complex regulatory element designated *Spm. Spm* consists of two components, *c1* (which is a suppressor element regulating *I*) and *c2* (which encodes a transposase that regulates transposition and changes in state of both *I* and *c1*).

When active, *c1* causes *I* to suppress expression of a structural gene (such as *A*) when *I* is inserted near *A*. When *c1* is inactive, *I,* in the same position, allows full or partial expression of *A,* depending on the position and state of *I*. Thus, *c1* and *I* act in conjunction to influence *A* expression. Changes in the state of *I* or *c1* (active or inactive) occur by mutation-like events regulated by *c2*. Based on this information, we can postulate that *c1* is capable not only of transposition, but also of inversion at a particular site in the corn genome, and that in only one of the two invertible orientations is *c1* active. The *c2* gene product may catalyze inversion as well as transposition, and imprecise recombination may cause changes in state of the *c1* or *I* sequences. It is also of interest that *c2* can catalyze its own transposition. In fact, all or part of the *Spm* element is mobile.

Sometimes *c1* is associated with a controlling element that does not respond to *c2*. Under these conditions, *c1* may show phase changes, reversibly going from the active to the inactive state, in response to developmental signals. This fact can give rise to differences of pigmentation within different tissues or during different stages of development in a single plant. As the relationships existing between these genes are exceptionally complex, it seems reasonable that they evolved to serve a function related to the developmental control of gene expression. In other words, paramutation in plants, controlled at least in some cases by transposable elements, may normally function as a mechanism for regulating gene expression during development. The same can be postulated for epigenetic phenomena in animals. The occurrence of mechanistically related events in microorganisms attests to the appearance of these processes prior to the development of multicellularity. It is in microorganisms, however, that the mechanistic details are best understood.

Switch Mechanisms in Microorganisms

Several examples of developmentally significant switch processes in microorganisms are considered in other chapters. Thus, a switch in flagellin synthesis occurs during the normal cell cycle of *Caulobacter* (Chapter 2); yeast cells are capable of interconverting mating types (Chapter 12); and the epigenetic control of antigen expression in *Paramecium* during aging has been well documented (Chapter 13). In *E. coli,* synthesis of many proteins occurs at discrete times in the cell cycle, while the same is true during sporulation in *Bacillus* and heterocyst formation in Cyanobacteria (Chapters 4 and 5). These gene activation and silencing processes comprise cell cycling and temporal differentiation programs, but the mechanisms are not understood.

Discrete switch processes have been observed in a variety of bacteria, including *Myxococcus, Salmonella,* and *Bacillus.* Representative well documented switches involving easily detectable cell surface antigens in Gram-negative bacteria are listed in Table 11.2.

Gram-negative bacteria possess numerous distinct species-specific antigens, all of which are localized within or external to the outer portion of the cell envelope (Table 11.2). Of these, the most extensively investigated are the O-antigens (lipopolysaccharides) and the H-antigens (flagella). In *Salmonella* species (of which there are many) both O- and H-antigens are subject to epigenetic (phase) variation. Only one antigen of a particular type can be expressed at any one time by a given cell, although that cell may possess the genetic potential to express two, three, or even four distinct antigens of that same type. Such cells are said to be bi-, tri-, or tetraphasic, respectively. In a biphasic *Salmonella* cell,

Table 11.2. Extracellular Macromolecular Structures in Gram-negative Bacteria that Exhibit Phase Variation

Structure	Transition	Function
Flagella	H1 ↔ H2	Motility
Lipopolysaccharide	O ↔ O′	Recognition
Fimbriae	+ ↔ ± ↔ −	Adhesion
Pili	Male ↔ Female	Conjugation
Capsular polysaccharide	− ↔ +	Protection

initiation of the expression of one H-antigen gene corresponds to cessation of expression of the other. Expression is apparently subject to intergenic repression and is mutually exclusive. In contrast to *Salmonella, E. coli* is monophasic with respect to H-antigen expression.

A specific determinant of the O-antigenic lipopolysaccharide (termed O-antigenic factor 12_2) in certain groups of *Salmonella* species undergoes form variation. It has been shown that this variation results from the presence or absence of the structural unit: O-α-D-glucopyranosyl (1 → 4) D-galactopyranose in the intact lipopolysaccharide. Moreover, the genetic unit responsible for form variation (for interconversion of the alternative phenotypes) maps on the *Salmonella* chromosome near the structural gene coding for a glucosyl transferase, which is responsible for the biosynthetic addition of terminal glucosyl residues to the chains of sugar residues in the lipopolysaccharide. It therefore appears that the presence or absence of the biosynthetic glucosyl transferase is responsible for O-antigen form variation in this instance. Form variation of bacterial cell surface antigens (H and O, for example) may serve as a defense mechanism against host immunity systems.

Fimbriae (Type I, subtype I) are, in part, responsible for the adhesive properties of some Gram-negative bacteria. Several hundred fimbrial filaments (7 nm in diameter, 0.2–20 μm in length) are peritrichously arranged over the surface of the bacterium. Each filament is composed of a helical array of hundreds of copies of a single protein subunit (fimbrilin, MW = 16,600) attached to a basal body in the cell envelope. Fimbrilin has been shown to be an agglutinin with binding specificity for α-D-mannopyranosyl residues in glycolipids and glycoproteins. By virtue of this adhesive organelle, Gram-negative bacteria can bind to the surfaces of a variety of bacterial, fungal, animal, and plant cells that possess mannose-containing cell surface molecules.

Production of fimbriae is due to expression of a chromosomal gene and is subject to phase variation. This variation appears to be more complex than biphasic flagellar phase variation (see below) or O-antigen form variation because fimbrial production can apparently be tuned up or down. It seems that *quantitative* variation in the amounts of fimbrilin synthesized is subject to control by a reversible process that is genetically unstable. This sort of *quantitative* phase variation is reminiscent of *R* factor control of aleurone pigmentation in maize.

Finally, other bacterial functions have been shown to change with high frequency. Other cell surface macromolecules, catabolic enzymes, colony morphology, and cell pigmentation have all been shown to yield to epigenetic variation in certain instances. In a few cases, pleiotropic switches occur. For example, certain fresh isolates of marine bacteria, which exhibit bioluminescence, can turn the lights on and off, apparently by a process analogous to phase variation, but correlated with a number of physiological changes. The bioluminescent form exhibits bright, creamy, raised colonies on an agar surface, while the nonluminous form forms flat, translucent colonies. This dark variant is also much less motile than the normal bioluminescent form and is resistant to a bacteriophage to which the normal variant is sensitive. Thus, reversible switches in bacteria may control a single function, or they may pleiotropically influence a number of cellular physiological properties.

Biphasic Flagellar Phase Variation in *Salmonella*

Of the bacterial switch processes described in the previous section, the best characterized is that involving the flagellar filaments (H-antigens). In *Salmonella typhimurium* strain LT-2, H-antigen expression oscillates between two distinct antigens, H1 and H2, owing to synthesis of two distinct flagellin proteins of different molecular weights and amino acid compositions that assemble into two different types of flagellar filaments. These two proteins are encoded by two separate flagellin structural genes on the *Salmonella* chromosome. The genetic elements involved in phase variation are shown in Figure 11.2, and these elements are defined in the figure legend.

While the *H1* gene maps in a region of the chromosome coding for other constituents of the bacterial flagellum, the *H2* gene maps distant from all other genes controlling cellular motility. The *H2* gene maps together with a promoter gene (*ah2*), a gene controlling phase variation (*vh2*), and a gene (*rh1*) which codes for a repressor of the *H1* structural gene. The *H1* gene is flanked on one side by a *cis*-acting positive regulatory gene, *ah1*, which is probably the transcriptional promoter of the *H1* cistron, and a *cis*-acting negative regulatory gene, *op1*, which functions as the operator to which the *rh1* repressor binds to shut off *H1* synthesis. The *H2* and *rh1* structural genes comprise an operon under *ah2* control. The *vh2* gene determines the frequency of the *H1* ↔ *H2* transition.

The *H1* and *H2* genes show considerable homology and are likely to be of a common evolutionary origin. The *H2* gene of *Salmonella* can be made to replace *H1* in either *E. coli* or in *Salmonella* species. When replacement occurs, all controlling elements in the *H2* region are permanently lost, and the *H2* gene (at the *H1* locus) becomes sensitive to repression by the *rh1* gene product. The *H2* gene probably arose from a primordial flagellin gene at the *H1* locus as a result of gene duplication, translocation, and evolutionary divergence. Tri- and tetraphasic strains probably arose by additional gene duplications of the *H1* (or *H2*) genetic regions during evolutionary history.

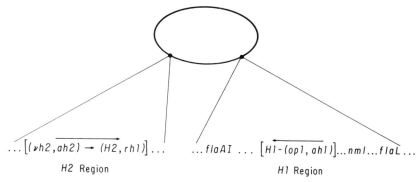

Figure 11.2. Genetics of the flagellar phase variation system in *Salmonella* species. *H1:* The structural gene for phase 1 flagellin. *H2:* The structural gene for phase 2 flagellin. *ah1:* Activator gene of the *H1* cistron; operates only *cis*; probably a promoter. *ah2:* Activator gene of the *H2* operon; operates only *cis*; probably a promoter. *rh1:* Structural gene for the *H1* repressor; part of the *H2* operon; operates both *cis* and *trans*—dominant in turning off *H1*. *rh1* does not operate on *H2*. *op1:* Operator gene of the *H1* cistron, to which the *H1* repressor must bind in order to turn off *H1* expression; operates *cis* but not *trans*. *vh2:* The gene that controls the rate at which H-antigen expression reversibly switches from one phase to the other; does not operate *trans* on another *H2* gene present in the cell. *nml, flaL,* and *flaAI* designate flagellar genes outside of the *H1* operon.

When the entire *H2* region in *Salmonella* is transferred into *E. coli* (which is normally monophasic), phase variation is observed. The *E. coli H1* gene is apparently subject to control by the *rh1* gene product. The frequency of phase variation in *E. coli* is higher than in the *Salmonella* genetic background. It therefore seems reasonable that since *E. coli* possesses the *op1* regulatory DNA, it must have evolved from an ancestral bacterium that exhibited phase variation but subsequently lost this trait because of its lack of survival value.

As a result of the application of genetic cloning techniques to the flagellar system, it has been possible to elucidate several important aspects of the mechanism of flagellar phase variation in *Salmonella*. Adjacent to, and to the left of, the *H2* structural gene is a genetic element that determines whether or not the operon consisting of the *H2* and *rh1* genes is transcribed. This DNA segment (of 980 base pairs) includes the promoter of the *H2* operon and can exist in either of two orientations, one in which the promoter is in the correct orientation, relative to the *H2* and *rh1* genes, for transcription of the operon, the other in which the 980 base pair sequence is in the opposite orientation so that the promoter cannot promote transcription of the *H2* operon. Inversion (switching) presumably results from intramolecular recombination between reverse repeated nucleotide sequences that flank the 980 base pair invertible region. Present within the invertible nucleotide sequence is a gene, designated *hin* (for "*H* *in*version") that codes for an enzyme, a "flipase", which specifically catalyzes the inversion process. When this gene is mutated so that the protein product becomes non-functional, the normal frequency of inversion (about 10^{-3} per cell per generation) is reduced to that expected of normal mutational rates (about 10^{-7} per cell per generation).

These facts were established following isolation (cloning) of the regulatory and structural regions of the *H2* DNA both in the "on" and "off" configurations. When the cloned DNA was transferred to an *E. coli* strain which was defective for the flagellin gene, phase variation was observed with the strain alternating between the *H2* on (motile) and *H2* off (nonmotile) states, thus showing that the entire regulatory region as well as the *H2* structural gene was intact in the cloned DNA fragments.

Heteroduplex analyses of the cloned DNA established that a 980 base pair segment of the DNA was inverted when the *H2* region was in the "on" configuration relative to the "off" configuration. In these experiments both "on" and "off" double-stranded DNA was "melted" to yield single strands, the two cloned populations were mixed, and they were allowed to reanneal. Electron microscopy revealed that adjacent to the H2 gene was a region of nonhomology that yielded a "bubble" owing to the presence of two single strands of nonhomologous DNA within the double-stranded homologous segments (Figure 11.3). This region corresponded to the *vh2* region, characterized previously by genetic techniques. The lack of homology was due to an inversion event that occurred whenever the operon switched from the "on" to the "off" configuration.

DNA sequence analysis revealed that the 980 base pair invertible region contains a large open reading frame, which presumably defines the *hin* gene, and that this region is flanked by reverse repeated DNA sequences of 14 base pairs each. These results and others established that the switch process involved inversion of a DNA fragment that was assumed to code for the *H2* promoter as well as the *hin* gene. The *hin* gene promoter is apparently present within the inversely repeated 14 base pair sequence at the two ends of the invertible region. Since this sequence is the same (but inverted) on the two sides, transcription of the *hin* gene is promoted regardless of the orientation of the DNA, and thus the flipase is always made. However, promotion of transcription of the *H2* operon (the *H2* and *rh1* genes) only occurs with the *H2* promoter adjacent to these structural elements as is observed when the *vh2* region is in the "on" configuration. These details are illustrated in Figure 11.4.

Inversion of the G Segment in Phage Mu

The bacteriophage Mu is capable of switching its bacterial host range by a mechanism that is analogous to flagellar phase variation. In one of the two possible states, termed ($+$), the phage is capable of infecting *Escherichia coli* K12 but none of a variety of other Gram-negative bacteria. Phage Mu is capable of switching its host specificity with a frequency of about 0.03 times per phage particle per generation. In the alternative state, termed ($-$), it is incapable of infecting *E. coli*, K12, but exhibits specificity toward *Citrobacter freundi* and certain other Gram-negative bacteria.

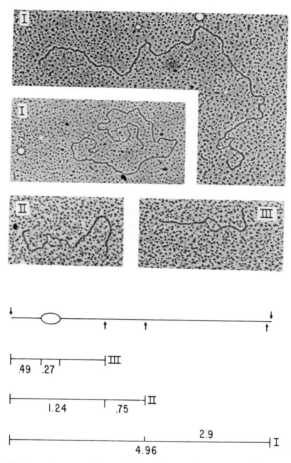

Figure 11.3. Electron microscopic analysis of heteroduplexes between cloned *H2* DNA in the "on" and "off" configurations. The micrographs reveal a region of nonhomology, 980 base pairs long within the cloned fragment. This region of nonhomology was shown to be due to inversion of the DNA sequence corresponding to the *vh2* region of the *H2* operon. I, II, and III show different restriction fragments, all of which contain the *vh2* region. Below is a map of the *H2* DNA indicating positions of restriction enzyme cleavage (shown by the arrows). Dimensions are given in microns. From J. Zieg, M. Silverman, M. Hilmen, and M. Simon, *Science 196:*170–172 (1977), copyright 1977 by the AAAS, reprinted with permission.

Like flagellar phase variation in *Salmonella,* host range variation results from inversion of a segment of the phage genome. In this case the invertible strand, termed the G segment, is 3000 base pairs long, representing about 8% of the total phage DNA. Inversion presumably results from intramolecular recombination between inversely repeated DNA sequences, 34 base pairs long, which flank the G segment. As for flagellar phase variation, a single enzyme, encoded

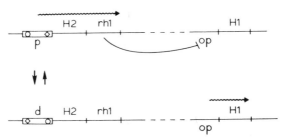

Figure 11.4. The proposed mechanism of flagellar phase variation in *Salmonella*. An inversion of the 980 base pair DNA sequence between the reverse repeated DNA sequences results in a change in orientation of the promoter of the *H2* operon (*P*). Only with *P* adjacent to the *H2* gene (top) is the *H2* operon transcribed. Recombination at the inverted repeats results in inversion of the *vh2* region so that *H2* genome expression is not initiated (bottom). Wavy line indicates transcription initiated from *P* located in the invertible region, indicated by the rectangles. *H1* and *H2* are the structural genes for the two flagellins; *rh1* is the structural gene for the *H1* repressor which interacts with the *op* site adjacent to the *H1* structural gene. From I. Herskowitz, L. Blair, D. Forbes, J. Hicks, Y. Kassir, P. Kushner, J. Rine, G. Sprague, Jr., and J. Strathern in *The Molecular Genetics of Development*, (T. Leighton and W.F. Loomis, eds.), Academic Press, 1980, reprinted with permission.

by the *gin* ("*G in*version") gene, catalyzes inversion of this segment and thus functions as the "flipase." The *gin* gene is adjacent to and to the right of the G segment (Figure 11.5).

To a first approximation, four genes are present on, or are partially encoded by the G loop. Two of these genes, denoted *S* and *U*, are read off of one DNA strand in one direction, while the other two genes, designated *S'* and *U'*, are read off of the other strand in the opposite direction. These genes code for the proteins that comprise the phage tail fibers and function in adsorption of the phage to a cell surface macromolecule, probably lipopolysaccharide in the host bacterium (see Chapter 10). These proteins confer upon the phage its host range specificity. The promoter is apparently outside and to one side of the G segment. The situation is therefore the converse of that of the *H2* operon where the promoter and inverting enzyme are within the invertible segment. Here these two regulatory elements map outside the G loop while the structural genes (*S* and *U*, or *S'* and *U'*) of the operons undergo inversion.

The actual situation is somewhat more complex. While the *U* and *U'* genes are fully encoded within the G loop, the *S* and *S'* genes are only partially included within the 3000 base pair invertible segment, and these variable parts of the genes are termed *Sv* and *Sv'*, respectively. A constant region of the *S* and *S'* genes is located outside of the G segment and adjacent to it. This common *S* region (*Sc*) is joined by G inversion to either one or the other of the two variable parts of the *S* gene (*Sv* or *Sv'*). The *S* or *S'* protein synthesized therefore represents the product of a genetic hybrid between the *Sc* region and either the *Sv* or the

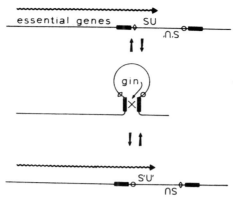

Figure 11.5. G region inversion in phage Mu. Transcription, indicated by a wavy line, is initiated from a promoter to the left of the G segment. The dark blocks represent the 34 DNA base pair inverted repeats in which recombination occurs and which flank the invertible G segment. The figure shows gene transcription in the + orientation (top), recombination (middle), and gene transcription in the − orientation (bottom). The X in the middle figure indicates intramolecular recombination. From I. Herskowitz, L. Blair, D. Forbes, J. Hicks, Y. Kassir, P. Kushner, J. Rine, G. Sprague, Jr., and J. Strathern in *The Molecular Genetics of Development,* (T. Leighton and W.F. Loomis, eds.), Academic Press, 1980, reprinted with permission.

Sv' gene, respectively. Clearly other variations upon this mechanistic theme are possible.

It is of considerable evolutionary interest to note that the G loop is found in another *E. coli* phage, P1. Moreover, the *hin* gene in *Salmonella* and the *gin* gene in phage Mu are 80% homologous, and these two flipases can substitute for one another catalytically. This suggests that these genetic elements may be or may once have been transposable as well as invertible, and that they arose from a common ancestral nucleotide sequence (an inversion sequence) that could insert into bacterial and phage chromosomes in a fairly nonspecific fashion. Possibly they inserted or were otherwise incorporated into eukaryotic DNA as well. The evolutionary advantage of these switch mechanisms to the bacteria and the phage are clear. Flagellar phase variation allows *Salmonella* to escape the host immune system, while G loop inversion allows the phage to expand its host range without loss of specificity. It is of interest to note that phase variation in viruses and microorganisms frequently influences the potential for essential interactions with environmental agents.

Invertible and Transposable Elements in Differentiation

In Chapter 4 it was suggested that one mechanism to effect a cyclic or linear program of differentiation was to utilize a series of "sigma factors" that confer operon specificity (promoter specificity) on RNA polymerase. A second possible mechanism for switching genes on and off involves DNA methylation; and a third type of switch mechanism, which can occur either reversibly or irreversibly, utilizes invertible DNA sequences as discussed in the preceding sections. Inversion of a DNA sequence can occur in a reversible fashion if the enzyme catalyzing inversion is continuously synthesized. This was the case for both the *hin* and the *gin* gene products because these structural genes and their promoters were intact and in proper orientation regardless of the orientation of the invertible DNA sequence. However the *hin* gene, present within the 980 base pair invertible segment of the *vh2* region, actually utilizes two promoters of very similar sequence because part of the *hin* promoter lies within the 14 base pair inverted repeat that flanks the two sides of the invertible segment. If the promoter for the *hin* gene had been outside of this region; if the promoter had been inside the invertible region with the structural gene outside this sequence, or if the *hin* gene was part of the H2 operon, inversion from the "on" orientation to the "off" configuration would comprise an irreversible event, because the flipase, catalyzing inversion, would be synthesized when the DNA was in only one orientation (see Chapter 14). It is therefore possible to envisage both reversible and irreversible genetic switches that are due to inversion of specific DNA sequences. If such a sequence encodes or functions to activate differentiation-specific proteins, a differentiation-specific switch in gene expression would result. Finally, if each such genetic rearrangement results in the silencing of the gene coding for the flipase for that sequence while activating a flipase for another invertible sequence, a linear irreversible program of differentiation could result as discussed in Chapter 14.

In the next chapter still a fourth mechanism that can lead to an irreversible switch in gene expression, involving sexual interconversion in yeast, will be examined. A fifth mechanism, apparently involving permanent loss of genetic material, results in activation of immunoglobulin synthesis in higher animals and may be involved in mating type determination in certain ciliated protozoans (Chapter 13). It is thus becoming apparent that several mechanisms of gene activation, some of which involve genetic rearrangement and some of which do not, were probably available to evolving, differentiating organisms. Further studies will undoubtedly reveal a multiplicity of genetic activation processes, many or all of which are utilized during expression of developmental programs in higher organisms.

Selected References

Amano, T., E. Richelson, and M. Nirenberg. Neurotransmitter synthesis by neuroblastoma clones, *Proc. Natl. Acad. Sci. USA* 69:258 (1972).

Brink, R.A. Paramutation, *Ann. Rev. Gen. 7:*129 (1973).

Coe, E.H., Jr. and M.G. Neuffer. "The Genetics of Corn" in *Corn and Corn Improvement* (G.F. Sprague, ed.), American Society of Agronomy, Inc., Madison, Wisconsin, 1977.

Cooke, J. Methylation and gene control, *Nature 296:*602–603 (1982).

Doerfler, W. DNA methylation and gene activity, *Ann. Rev. Biochem. 52:*93–124 (1983).

Fedoroff, N.V. Transposable genetic elements in maize, *Sci. Amer. 250:* 85–98 (1984).

Fincham, J.R.S. and G.R.K. Sastry. Controlling elements in maize, *Ann. Rev. Gen. 8:*15 (1974).

Giphart-Gassler, M., R.H.A. Plasterk, and P. van de Putte. G inversion in bacteriophage Mu: a novel way of gene splicing, *Nature 297:*339 (1982).

Herskowitz, I., L. Blair, D. Forbes, J. Hicks, Y. Kassir, P. Kushner, J. Rine, G. Sprague, Jr., and J. Strathern. "Control of Cell Type in the Yeast, *Saccharomyces cerevisiae* and a Hypothesis for Development in Higher Eukaryotes" in *The Molecular Genetics of Development* (T. Leighton and W.F. Loomis, eds.), Academic Press, New York, 1980.

Jones, G.W. and R. E. Isaacson. Bacterial adhesions and their receptors *in CRC Crit. Rev. Micro. 10:*229–260 (1983).

Malawista, S.E. and M.C. Weiss. Expression of differentiated functions in hepatoma cell hybrids: high frequency of induction of mouse albumin production in rat hepatoma-mouse lymphoblast hybrids, *Proc. Natl. Acad. Sci. USA 71:*927 (1974).

Shay, J.W. 1983. Cytoplasmic modification of nuclear gene expression. *Mol. Cell. Biochem. 57:*17–26.

Silverman, M. and M. Simon. "Phase Variation and Related Systems" in *Mobile Elements* (J.A. Shapiro, ed.), Academic Press, New York, 1983.

Zieg, J., M. Silverman, M. Hilmen, M. Simon. Recombinational switch for gene expression, *Science 196:*170 (1977).

CHAPTER 12

Sex Determination and the Interconversion of Mating Type

We do not even in the least know the final cause of sexuality; why new beings should be produced by the union of the two sexual elements, instead of by a process of parthenogenesis . . . The whole subject is as yet hidden in darkness.

Charles Darwin

It has long been thought that sexual determination is an exceptionally complex, possibly incomprehensible process. This was particularly true of higher mammals including humans, where an organism may possess the karyotype, say of a male, but exhibit some of the characteristics of the female. Genetic disorders and artificial hormone therapy were known to give rise to phenotypic sex expression that did not reflect the sex chromosomal composition of the organism. Some degree of physical and psychological androgeny in humans is virtually universal, and in extreme cases, individuals exhibit physical characteristics of one sex but psychological and behavioral characteristics of the opposite sex. These observations emphasized the similarities and the extensive overlap that can occur between the male and female sexes. Every individual, regardless of sex, may possess the genetic potential to express the full complement of characteristics that typify both sexes.

In view of these observations with higher organisms, recent studies of sexual determination in lower organisms are of particular interest. In this chapter it will be seen that for yeast, the one organism where sex determination, expression, and interconversion are fairly well understood at the molecular level, the genome does, in fact, carry all of the genetic information necessary for full phenotypic expression of both sex types. Moreover, one genetic locus, *MAT,* consisting of only a very few regulatory genes (probably four) distinguishes the two sex types. These can be modified to yield partial sex phenotypes. Equally important in yeast are alleles that determine sex-dependent and sex-independent fertility, and the expression of these genes may be controlled by the four regulatory genes at the *MAT* locus. Finally, the potential for sexual interconversion is controlled by just a few (probably three) genes. Sexuality and its interconversion are therefore regulated by a very few gene products. While this chapter will deal largely with sexual determination in yeast, further interesting examples of sexual determination and interconversion are considered in the last two sections of the chapter and in Chapter 13. The degree to which the findings with the unicellular organisms examined here are applicable to higher animals has yet to be ascertained.

Sex Determination in Yeast

In many species of yeast, including *Saccharomyces cerevisiae,* alleles at a single genetic locus, termed the mating type locus (*MAT*) determine the sex of a particular haploid cell. A haploid yeast cell may be either of one sex, the α mating type, or of the opposite sex, the **a** mating type. A cell will be of the α mating type if it bears the *MAT*α allele, but of the **a** mating type if it carries the *MAT***a** allele. Only haploid cells of opposite mating type agglutinate and fuse resulting in the formation of a diploid cell of α/**a** genotype.

Several biochemical and physiological characteristics distinguish α from **a** cell types. First, as indicated in Chapter 9, α cells produce an oligopeptide pheromone, called "α-factor," which causes **a** cells to arrest in the G1 phase of the cell cycle. Subsequently, the arrested cells undergo morphological changes. α-Factor is cell type specific, being without effect on α cells or diploids. **a** Cells reciprocally produce a pheromone that acts selectively upon α cells. This pheromone, termed "**a**-factor," induces corresponding changes specifically in α cells (see Chapter 9). Second, α cells produce a cell surface mannan-protein complex that allows these haploid cells to undergo sexual agglutination with **a** cells although no agglutination reaction occurs with other α cells or with diploids. Similarly, **a** cells produce a glycoprotein that mediates complementary and specific recognition of the α cell surface glycoprotein. These two sexual agglutinins are clearly distinct molecules (see Chapter 10). Third, **a** cells (but not α cells) possess the ability to inactivate α-factor and subsequently to recover from cell cycle arrest. Finally, the genetic information encoded at the *MAT***a** and *MAT*α loci controls other cell type-specific processes such as meiosis and ascospore formation in diploids. These processes require the involvement of dozens if not hundreds of genes that are not essential for vegetative asexual growth. The expression of numerous biochemical attributes, encoded by genes that map throughout the yeast genome, is therefore controlled by genes present at the mating type locus, *MAT*.

The nature of the genes and gene products included within and encoded by the *MAT* alleles has not been fully elucidated. However, the *MAT*α and *MAT***a** alleles have been cloned, and heteroduplex studies have revealed that the two mating type loci show two homologous segments that flank a stretch of completely nonhomologous DNA. This stretch of nonhomologous DNA is 642 base pairs long in *MAT***a** (indicated by the vertical bars in Figure 12.1) and 747 base pairs long in *MAT*α (indicated by the diagonal bars in Figure 12.1). Further, each *MAT* locus has been shown to produce two discrete mRNA species, both of which are initiated within the regions of nonhomology and apparently code for separate and distinct regulatory proteins. One mRNA species from each *MAT* allele is synthesized rightward and is confined largely to the region of nonhom-

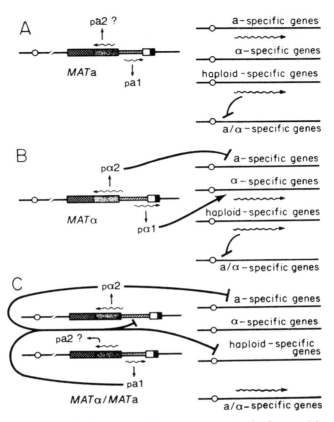

Figure 12.1. Control of cell type-specific gene expression in *S. cerevisiae*. Negative control ("turn-off") is indicated by a T-bar; positive control ("turn-on") is indicated by an arrowhead. Wavy lines represent poly A+ mRNAs. A large body of genetic and biochemical evidence supports the following model for regulation of cell type-specific gene expression by *MAT*. The **a** cell-specific functions are expressed constitutively (A). In an α cell, therefore, expression of **a** cell-specific genes must be prevented and, in addition, α cell-specific functions must become turned on. The protein product (pα2) coded for by the α2 transcript from *MAT*α is a "repressor" of **a** functions, and the product (pα1) of the α*1* transcript is an "inducer" of α functions (B). Certain functions are required by both haploid cell types, including products responsible for preventing expression of genes required for the sporulation process. In an **a**/α diploid, the presence of the product (pa1) of the **a**1 transcript from *MAT*a acts in concert with pα2 (perhaps by forming a protein–protein complex) such that it represses expression of all haploid–specific functions, including α*1* transcription, and consequently inhibits the ability of the cell to mate and also permits the expression of those functions necessary for its ability to undergo meiosis, sporulation, and other diploid cell-specific processes (C). The product (pa2) of the **a**2 transcript from *MAT*a may function in the induced synthesis of the α-cell-specific mating type agglutinin, but this proposal has not been completely established. The DNA sequences specific to α cells and to **a** cells are indicated by the diagonal and vertical bars, respectively. Other regions of the DNA are shared by α and **a** cells. From G.F. Sprague, Jr., L.C. Blair, and J. Thorner, *Ann. Rev. Micro. 37*:623–660 (1983). Reproduced with permission, from the Annual Review of Microbiology, Vol. 37. © 1983 by Annual Review Inc.

ology (the *a1* transcript, encoded within *MAT*a, and the α*1* transcript, encoded within *MAT*α). The other mRNA species from each *MAT* allele is synthesized leftward and is mainly encoded by an adjacent region of homology common to *MAT*a and *MAT*α (the *a2* transcript, encoded within *MAT*a, and the α*2* transcript, encoded within *MAT*α) (see Figure 12.1). The two proteins encoded at *MAT*α (designated pα1 and pα2) and those encoded at *MAT*a (designated pa1 and pa2) apparently control the expression of all sex-specific and sporulation-specific genes that map elsewhere in the yeast genome as revealed by mutational analyses.

Numerous yeast mutants defective in mating have been isolated and characterized. Mutations affecting mating fall into three classes: those causing defective mating only in α cells, only in **a** cells, or in both cell types. Of the first class, the α-specific mutations cause defects in α-factor production, in **a**-factor response, in agglutinin production, or in a subsequent step in the mating process (see Chapters 9 and 10). Similar mutants defective for mating responses of **a** cells have been isolated, and a few haploid-specific genes, expressed in both **a** and α cells but not in diploids, have also been identified. The vast majority of the mutations in these genes map at genetic loci other than within *MAT*, but a few, mapping within the *MAT* locus, have also been characterized.

Only two mutant phenotypes have resulted from defects that map within *MAT*α. These mutations, giving rise to the *mat*α*1* and *mat*α*2* mutant alleles, are both recessive to the wild type alleles, and they complement each other. These observations are consistent with the fact that two mRNA transcripts are encoded by the *MAT*α locus, and they lead to the conclusion that *MAT*α contains two structural genes that code for two regulatory proteins.

The properties of *mat*α*1* mutants are as follows: (a) they do not mate with either α or **a** cells; (b) they do not respond to **a**-factor; and (c) they do not synthesize active α-factor. *mat*α*1*/*MAT*a diploids do sporulate, however. On the other hand, *mat*α*2* mutants behave somewhat like **a** cells. They mate as **a** cells with α cells, but with only 1% of the efficiency observed with normal cells. Also, most of these mutants respond to and degrade α-factor, and they produce **a**-factor and the **a**-cell specific agglutinin. *mat*α*2*/*MAT*a diploids do not sporulate. On the basis of these results, it has been concluded that the *MAT*α locus codes for two regulatory proteins, pα1 and pα2, which respectively function as an activator of the α-specific genes and a repressor of the **a**-specific genes that map elsewhere on the yeast genome. Consequently, **a**-specific mating functions may be constitutively expressed in the absence of pα2 function while expression of α-specific mating functions requires the presence of a functional pα1 protein. A scheme illustrating this proposal is shown in Figures 12.1A and B.

Diploid yeast cells are capable of carrying out meiosis in a process wherein each diploid cell gives rise to an ascus containing four haploid spores. Because sporulation requires the α/**a** diploid state, and therefore requires expression of genetic information at both the *MAT*α and the *MAT*a loci, it has been proposed that each of these two loci codes for a sporulation-specific codominant activator protein (positive control) or a codominant repressor of haploid-specific proteins that functions to repress synthesis of the sporulation genes (negative control).

Indeed, either the mutant *matα2* allele or the mutant *mata1* allele (resulting from a mutation within *MAT*a) gives rise to loss of the normal sporulation activity of the α/**a** diploid. It is therefore reasonable to suggest that the multitude of sporulation-specific (*spo*) genes in yeast are subject to control by two codominant regulatory proteins encoded by the *MAT*α2 and *MAT*a1 genes.

The second presumed regulatory gene within *MAT*a is the *MAT*a2 gene. Defects in this gene do not prevent mating but appear to cause constitutive expression of the α-cell-specific mating type agglutinin that is found on the surface of **a** cells (see Chapter 10). Normally the agglutinin is synthesized only when **a** cells are exposed to α-factor, but a defect in the *MAT*a2 gene (giving rise to the *mata2* allele) renders its synthesis independent of α-factor and therefore constitutive. Mechanistically, the pa2 protein, the product of the *MAT*a2 gene, can be thought of as a transcriptional repressor of α-cell-specific agglutinin synthesis, and this repressor may be active only when cyclic AMP levels are high.

Finally, it should be noted that diploid cells do not ordinarily produce α-factor, **a**-factor, the sexual agglutination-specific glycoproteins or other haploid-specific functions. In other words, sex-specific functions expressed in the haploid state are turned off by the combined action of *MAT*a and *MAT*α in α/**a** diploids. In order to account for this observation, it has been suggested that the a1 and α2 proteins may repress expression of all haploid-specific genes, and that among these genes are repressors of the diploid-specific (i.e., sporulation and meiosis-specific) genes. This postulate, which has not been established, suggests that the mechanism of activation of sporulation in the diploid state involves the prevention of a repressive mechanism. These proposals are depicted schematically in Figure 12.1C.

Mating Type Interconversion in Yeast

Sex changes interconverting females and males, as a result of mutation or an epigenetic change, are common in simple and complex eukaryotes. Only in one organism, however, *Saccharomyces cerevisiae,* has the molecular mechanism been clarified. Consequently, we shall pursue the phenomenon in this organism to the extent that present-day knowledge permits. Our first discussions will deal with the physiological consequences of mating type interconversion and will define the phenomenon; then a reasonable mechanism and the evidence favoring this mechanism will be presented. In the subsequent section, we shall discuss the rules that determine the propensity of a cell type to switch.

In yeast species, four different life styles may be encountered (Figure 12.2). First, a strain may be *heterothallic* (Figure 12.2A). Asci of a heterothallic (heterosexual) strain contain two stable spores of the α mating type and two of the **a** mating type. All four of these spores can germinate independently to produce haploid clones of identical cells. Within any one clone, all of the cells are of the same mating type, the mating type of the spore of origin. Diploidization occurs only if two haploid cells of opposite mating type are brought together

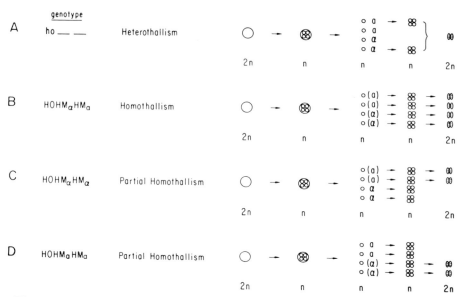

Figure 12.2. Four life styles or modes of mating exhibited by *Saccharomyces cerevisiae* and other yeast species. The genotypes of the four different strains are shown on the left.

(Chapter 10). Second, a yeast strain may exhibit *complete homothallic behavior* (Figure 12.2B). Each of the four ascospores within an ascus will germinate and divide with the formation of two haploid daughter cells. The second cell division gives rise to four haploid cells, which can (but need not necessarily) diploidize. This "homosexual" mating behavior *appears* to result from mating between like cells, but we shall soon see that this is not true. Finally, two types of *partially homothallic* strains have been observed. In the first type, each ascus produces two stable spores of the α mating type and two spores capable upon germination of diploidization (Figure 12.2C). In the second type, each ascus contains two stable spores of the **a** mating type as well as two spores that will undergo selfing if the daughter cells are not separated (Figure 12.2D).

It is possible to analyze the mating type of a germinating spore or a vegetative cell in either a heterothallic strain or a homothallic strain, employing haploid, heterothallic "tester" strains of known mating type. Such analyses reveal that even in homothallic strains, two of the four ascospores initially are of the α mating type while the other two are of the **a** mating type. In heterothallic strains these are stable, but in fully homothallic strains, cells within the clone can switch with a certain probability to the opposite mating type. In partially homothallic strains, one mating type, say α, can switch to **a**, but the reverse transition cannot occur. It is therefore clear that, while complete homothallism comprises a reversible switch in mating type, partial homothallism represents an irreversible switch. The molecular basis of these differences will be explained below.

Genetic analyses of yeast strains exhibiting the four modes of sexual behavior have shown that, in addition to the mating type locus, *MAT*, three genetic loci each with two alternative alleles, determine the ability of a haploid clone to undergo mating type interconversion. First, the dominant *HO* gene is required for either complete or partial homothallism. When the recessive allele, *ho*, is present, the strain is heterothallic. The *HO* gene is consequently the *master switch*, which allows for the switching mechanism. It is located on chromosome IV and probably codes for the enzyme (an endonuclease) that initiates mating type interconversion. The second and third genetic loci are referred to as *HML* (homothallic mating gene on the left side of chromosome III) and *HMR* (homothallic mating gene on the right side of chromosome III). The reason for this genetic nomenclature is shown in Figure 12.3. All three genetic loci, *HML* (left), *MAT* (center), and *HMR* (right) are localized to a single chromosome, chromosome III of the yeast genome. Moreover, each locus can accommodate either of two alleles, the α allele or the **a** allele. If *MAT* accommodates the α allele, (*MAT*α), the mating type will be α, but if the *MAT***a** allele is present, the haploid yeast cell will express the **a** mating type. If either the *HML* or the *HMR* locus exhibits the α allele (*HML*α or *HMR*α, respectively), a haploid strain of the **a** mating type will, in the presence of the dominant *HO* allele, be able to switch mating type from **a** to α, but in order for the α to **a** transition to occur, either the *HML***a** or the *HMR***a** allele must also be present. Hence, a strain possessing the *ho* allele will be heterothallic, regardless of the genetic composition of the *HM* loci (Figure 12.2A), and a *HO HML*α *HMR***a** strain will exhibit complete homothallism (Figure 12.2B). In a strain of the *HO HML***a** *HMR***a** genotype, only the α→**a** transition will occur (Figure 12.2C), while a strain of the *HO HML*α *HMR*α genotype can only undergo the **a**→α transition.

Early investigators of mating type interconversion considered a mechanism involving inversion of a segment of DNA within or adjacent to *MAT*. When this

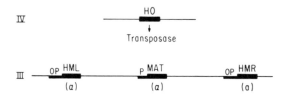

Figure 12.3. Genetic structures of genes involved in mating type interconversion in *S. cerevisiae*. The *HML, MAT,* and *HMR* loci are localized to chromosome III and can exist in either of two allelic forms, the α or the **a** form. The *HO* gene is localized to chromosome IV and also can exist in either of two allelic forms: The *HO* allele is required for mating type interconversion in either direction while the *ho* allele is insufficient for (cannot initiate) mating type interconversion. Thus, *HO* can be thought of as the "catalyst" or "master switch"; whereas, *HM*α "passively" permits the **a** → α transition to occur, and *HM***a** "passively" allows the α → **a** transition to occur. Mechanistically, the *HO* gene product may be a sequence-specific nuclease or "transposase" while *HM*α and *HM***a** code for silent copies of the mating type information which, when transposed to *MAT*, become activated.

putative DNA segment was inserted left to right, the α mating type would be expressed. When it was inserted right to left, the **a** mating type would be expressed. This mechanism would be analogous to phase variation in *Salmonella* or to G inversion in phage Mu (see Chapter 11). The first evidence that such a mechanism was *not* involved came from "healing" experiments. A mutation was introduced at the mating type locus; i.e., *matα1* replaced the wild type allele, *MATα*, in a strain exhibiting complete homothallism. When the strain had switched from the α mating type to the **a** mating type and then back to α, the wild type allele, *MATα*, replaced the defective *matα1* gene. Under no conditions could the mutant allele be recovered. It was apparently lost. Further, the same experiment could be repeated with other mutant alleles at *MAT* (i.e., employing mutants in the *matα2 or mata1* genes) with the same result. These observations led to the unavoidable conclusion that the genome must possess "silent" copies of *MATa* and *MATα*, and that these must be mobile in the sense that this silent genetic information could be copied or transferred to the *MAT* locus, thereby replacing the pre-existing genetic information and becoming activated.

These observations and others led to the *cassette* model which consisted of several interrelated postulates: (1) *MAT* DNA (either *MATα* or *MATa*) is like a cassette tape which is expressed if and only if it is "plugged in" at the *MAT* locus. (2) The promoter within *MAT* is like a playback head of a tape recorder and allows transcription of the included gene. (3) Silent *MAT* copies exist at the *HML* and *HMR* loci, and either *MATα* or *MATa* DNA can be present at either of these two sites. (4) Expression of the *MAT* DNA at the *HMR* and *HML* sites is prevented because of a repressor mechanism that is operative at *HML* and *HMR* but not at the *MAT* locus. (5) Interconversion (replacement of *MAT* with a silent copy from *HML* or *HMR*) requires the participation of the *HO* gene, and replacement results in loss of the genetic material originally at *MAT*. (6) The *HM* loci remain unaltered during and after switching.

A number of independent experiments now support this model. First, introduction of a mutation at either *HMLα* or *HMRa* results in switching between a wild type allele (**a** or α) and a mutant allele (α or **a**), respectively. Second, the four genes, *HMLα*, *MATα*, *HMRa*, and *MATa* have been cloned and examined in heteroduplex studies. The results showed that *HMLα* and *MATα* possess a 747 base pair region of homology which they do not share with *HMRa* or *MATa*. These latter two genes, however, possess a 642 base pair region of homology which is not found in *MATα* or *HMLα*. Third, interconversion of mating type is observed at a very low frequency (10^{-7} events per cell per generation) in an *ho* genetic background. Interconversion in the absence of the *HO* allele can apparently occur by chromosomal rearrangement in a process that results in fusion of the *HML* or *HMR* genetic region to homologous regions within the *MAT* locus. This observation provides convincing evidence that *HMLα* and *MATα*, and *HMRa* and *MATa* are functionally identical. Fourth, "suppressor" mutations allow heterothallic mutant strains that are defective at the mating type locus to mate and sporulate. For example, the *sir1* mutation (allelic with the wild type *SIR* [silent information regulator] gene) partially corrects *matα1*, *matα2*, *and mata1* defects, allowing low level mating with the haploid cell of opposite mating type and sporulation in the diploid. Sporulation in *sir1/sir1* diploids occurs even in *MATa/MATa* or *MATα/MATα* diploids. *sir1* is recessive to the wild type *SIR* allele.

In order to explain these observations, it has been suggested that the mutant *sir1* allele allows *low level* expression of *HMLα* and *HMRa* because *MATa sir1* and *MATα sir1* strains can mate, although with poor efficiency. Full expression of the *HM* loci should give rise to nonmating behavior as is observed for α/a diploids. It should be noted that other mutations are known that allow expression of *HMLα* and *HMRa*. These mutations define four complementation groups, suggesting that silencing of the *HM* loci is a complex process.

A proposal suggesting that the *SIR* genes encode a complex of DNA-binding proteins that act at the *HM* loci but not the *MAT* locus is illustrated in Figure 12.4. It is suggested that some region of the DNA (indicated by "O") is responsible for the binding of the *SIR* protein complex to the HM loci, and that the "O" region is lacking at *MAT*. The *SIR* protein presumably functions like a repressor of gene expression.

Pattern Control of Mating Type Switching

Careful examination of the mating capabilities of homothallic yeast cells during clonal growth has revealed that two general rules suffice to account for the switching pattern: (1) Both sister cells of any cell division are of the same mating type (i.e., α → 2α or 2a; a → 2a or 2α). From this observation, it must be concluded that mating type interconversion cannot occur continuously throughout the cell cycle. Instead it must occur between G1 and early S, before the *MAT* gene is replicated. (2) Only cells that have budded at least once (experienced cells that bear a bud scar) are competent to switch (Figure 12.5). Experienced cells switch mating type 73% of the time they undergo mitosis, while inexperienced cells switch less than 0.1%. Consequently, a clone of homothallic haploid

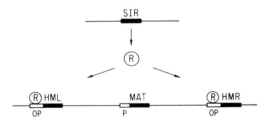

Figure 12.4. Proposed mechanism for the control of the *HM* loci by the four *SIR* (silent information-regulator) genes. The *SIR* genes are presumed to code for repressors (R), which bind to the "operator" (O) of the *HM* loci. Because there are no corresponding "operator" sites at the *MAT* locus, the mating type information at *MAT* is transcribed.

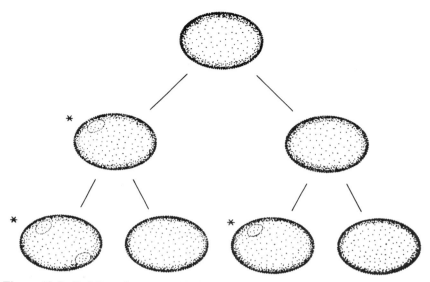

Figure 12.5. Budding of a yeast cell gives rise to two unequal cells, an "experienced" parental cell, bearing a bud scar, and a daughter cell, arising from the new bud. Only experienced cells are capable of switching mating type as indicated by the star. Since switching occurs only during the G1 phase of the cell cycle, both of the two daughter cells that result from cell division must be the same mating type.

cells contains cells that differ with respect to their competence to switch. It has been shown that the switching capability is retained for at least 12 generations of haploid growth, and is probably retained indefinitely. Diploidization, however, shuts off the switching mechanism altogether so that the $\alpha/\mathbf{a}$ diploid state is maintained stably.

A key to an understanding of these observations is based on the control of the expression of the *HO* gene. This gene appears to encode an endonuclease that specifically cleaves *MAT* DNA at the junction between the homologous and nonhomologous regions of the locus, thereby initiating mating type interconversion. Other enzymes, additionally required for mitotic and meiotic recombination, function together with the *HO* gene endonuclease to complete the process. Analyses of synchronized yeast cultures have shown that the *HO* gene is transcribed, and the encoded endonuclease appears only during the G1 phase of the cell cycle. Moreover, it is not found in inexperienced cells. How the cell-cycle dependent expression of this gene is controlled, and the factors determining its cellular distribution function have yet to be ascertained.

It should be noted that the asymmetry of switching capacity among progeny of dividing haploid homothallic yeast cells is reminiscent of the control pattern in heterocyst development in Cyanobacteria (Chapter 5). In the latter organisms, cell division gives rise to two unequal daughter cells, one smaller than the other, and the heterocyst *must* develop from the smaller cell. Possibly some *competence factor* is asymmetrically distributed between the two daughter cells prior to septum formation, and this factor, of unknown nature and function, controls the propensity to undergo a change.

Sex Determination in *Drosophila*

Recent work with *Drosophila melanogaster* has shown that in this organism, sex determination is subject to control by the X chromosome:autosome ratio and also by four major autosomal genes. These four genes may be complex regulatory loci that control the batteries of structural genes required for the development of many if not all secondary sex characteristics.

Two of these genes (designated *tra-1* and *tra-2* for transformer) when homozygous in chromosomal females give rise to phenotypic males with normal male external structures, internal genital tract, and courtship behavior. The third gene (designated *dsx* for doublesex) when homozygous in either chromosomal males or females causes the organisms to develop as hermaphrodites. This locus may be functionally complex, specifying two alternative functions in a mutually exclusive fashion. In males it may specify a product that acts to preclude expression of genes involved in female sexual differentiation, while in females it may specify a different product that precludes expression of male-specific genes. The fourth gene (called *ix* for intersex) transforms females but not males into hermaphrodites. These four genes apparently regulate sexual development in all tissues other than the germ line.

Results obtained in studies of these four mutant genes have shown that they exhibit a simple pattern of epistasis. Thus, the *dsx* gene is epistatic over the *tra* genes, which in turn are epistatic over *ix*. In other words, *ix* gene expression apparently depends on expression of the *tra* genes, and *tra* gene expression depends on *dsx*. These observations suggest that all four genes function within a single pathway that determines the sex of an organism. In this respect *Drosophila* resembles *Saccharomyces,* where the four principal regulatory genes encoded at the *MATa* and *MATα* loci function together to control expression of all sexual characteristics of the organism. Evidence is available to suggest that analogous regulatory genes exist in mammalian species including humans. The apparent genetic simplicity of sexual determination seems to be characteristic of higher organisms as well as unicellular organisms such as yeast.

It is worth noting that sex determination in eukaryotes provides a dramatic example of developmental control of gene expression. Sex determination must be considered as a major component of the organismal developmental program. Studies of sex determination such as those described in this chapter should lead not only to an understanding of this process, *per se,* but also to elucidation of the mechanisms effecting the expression of alternative developmental pathways.

Sexual Interconversion in Higher Plants and Animals

Higher plants and animals exhibit sexual variations indicating that in these organisms, as in single-celled microbes, only a few genes determine and control sexuality. Thus, within a particular genus, both hermaphroditic species, in which both the female and male sexual organs are found in a single organism, and bisexual species, in which female and male organs are found in separate organisms, are common. Lower animals such as *Hydra,* many plants, and a few genera of higher animals exhibit both bisexual and hermaphroditic life styles. Moreover, many organisms, including a variety of fish species, show sexual interconversion (i.e., female to male conversion), either as a normal part of a developmental program or as an occasional event. Phenotypic hermaphroditism has been reported even in humans.

In plants, true hermophrodites are those in which both female and male sexual organs are present within a single flower. Species in which flowers of the two sexes are distinct, but present on the same plant, are called monoecious, while those in which separate organisms produce pollen and eggs are referred to as dioecious. Sexual development in plants may be genetically or developmentally controlled; in the latter case, environmental and hormonal factors may be important.

Within each major classification of sexual behavior one finds considerable variation. A hermaphroditic species may be capable of self-fertilization, or self-sterility alleles may prevent, or partially inhibit, selfing. In some hermaphroditic organisms, self-conjugation is physically impossible while in others sperm and eggs are produced sequentially so that self-fertilization is minimized.

The genetic examination of sexual variations in higher organisms has led to the conclusion that genes for self-sterility, for sex determination, for sexual interconversion, or for hermaphroditism may be polymorphic within a given species. Selection of alternative alleles may determine the mode of reproduction employed by an organism, as well as the rate of interconversion of the available sexual forms. It must therefore be concluded that fertility is an arbitrary choice for delineating a species.

If different modes of sexuality are widespread among higher organisms, and if these forms can interconvert depending on the presence or absence of one or a few alleles, there must be evolutionary pressure (survival value) for the maintenance of each mode. Hermaphroditism appears to be favored whenever the population density is low or when a single organism is colonizing a new territory. Under these conditions, the probability that a mate will be available is low, and selfing is the favored mode of reproduction. Bisexuality (reproduction involving two sexes) is favored whenever outbred mates are available for three reasons: First, inbred offspring, avoided by outbreeding, are usually of low fitness owing to the expression of deleterious recessive alleles. Second, two separate sexes allow the most fit male to mate with the most fit female; and third, separate sexes provide maximal opportunity for DNA recombination, and therefore for evolutionary change. This last factor has only long-term survival value, but the first two considerations provide short-term advantages to the species. These considerations explain why different modes of reproduction, involving outbreeding as well as selfing, are each of survival value under appropriate environmental conditions, and therefore why an organism may maintain polymorphic alleles controlling different reproductive modes and their interconversion.

Selected References

Baker, B.S. and J.M. Belote, Sex determination and dosage compensation in *Drosophila Melanogaster, Ann. Rev. Genet. 17:*345–393 (1983).

Baker, B.S. and K.A. Ridge. Sex and the single cell. I. On the action of major loci affecting sex determination in *Drosophila melanogaster, Genetics 94:*383 (1980).

Bell, G. *The Masterpiece of Nature* The Evolution of Genetics and Sexuality, University of California Press, Berkeley, 1982.

Goodenough, U.W. and J. Thorner. "Sexual Differentiation and Mating Strategies in the Yeast *Saccharomyces* and in the Green Alga *Chlamydomonas*" in *Cell Interactions and Development: Molecular Mechanisms* (K. Yamada, ed.), John Wiley & Sons, Inc., New York, 1982.

Herskowitz, I., L. Blair, D. Forbes, J. Hicks, Y. Kassir, P. Kushner, J. Rine, G. Sprague, Jr., and J. Strathern. "Control of Cell Type in the Yeast *Saccharomyces cerevisiae* and a Hypothesis for Development in Higher Eukaryotes" in *Microbial Differentiation* (T. Leighton and W.F. Loomis, eds.), Academic Press, New York, 1980.

Klar, A.J.S., J.N. Strathern, J.R. Broach, and J.B. Hicks. Regulation of transcription in expressed and unexpressed mating type cassettes of yeast, *Nature 289:*239 (1981).

Klar, A.J.S., J.B. Hicks, and J.N. Strathern. Directionality of yeast mating-type interconversion, *Cell 28:*551 (1982).

Leighton, T. and W.F. Loomis, eds. *The Molecular Genetics of Development,* Academic Press, New York, 1980.

Matsumoto, K. I.Uno, Y. Oshima, and T. Ishikawa. Isolation and characterization of yeast mutants deficient in adenylate cyclase and cAMP-dependent protein kinase, *Proc. Natl. Acad. Sci. 79:*2355 (1982).

Naftolin, F. Understanding the bases of sex differences, *Science 211:*1263 (1981).

Nasmyth, K.A., K. Tatchell, B.D. Hall, C. Astell, and M. Smith. A position effect in the control of transcription at yeast mating type loci, *Nature 289:*244 (1981).

Oshima, Y. and I. Takano. Genetic controlling system for homothallism and a novel method for breeding triploid cells in *Saccharomyces, Proc. IV IFS: Ferment. Technol. Today* 847–852 (1972).

Sprague, G.F., Jr., J. Rine, and I. Herskowitz. Homology and non-homology at the yeast mating type locus, *Nature 289:*250 (1981).

Sprague, G.F., Jr., L.C. Blair and J. Thorner. Cell interactions and regulation of cell type in the yeast *Saccharomyces cerevisiae, Ann. Rev. Micro. 37:*623–660 (1983).

CHAPTER 13

Genetic Control of Development, Mating Type Determination, and Programmed Death in Ciliated Protozoa

Death belongs to life as birth does,
The walk is in the raising of the foot as in
The laying of it down.

Tagore

Death is life, and life is death;
They are locked together in an
Eternal, mad love-battle.

Hesse

The borders which divide life
and death are at best vague.

Poe

For that which is born, death is certain,
and for the dead, birth is certain.

The Bhagavad-Gita

Life without death is incomplete.

Pasteur

Somatic cells within a multicellular organism are linearly programmed to differentiate and die either during embryological development, or subsequently after maturation of the organism (Chapter 2). This fact necessitated, during evolutionary history, the separation of the soma from the germ line as well as the establishment of sexuality to provide organismal rejuvenation and species continuity (Chapter 3). Similar processes of differentiation, sex, and programmed death have evolved in the unicellular ciliated protozoa, but because the entire program of differentiation occurs within a single cell, somewhat different mechanisms and structural elements must underlie these processes. This chapter deals primarily with development in the ciliates. These organisms provide the simplest microbial system in which programmed death is known to occur as the terminal step in a sequence of developmental events.

Genetics of *Paramecium* and *Tetrahymena*

The ciliated protozoa such as *Paramecium* and *Tetrahymena* represent an interesting evolutionary *cul-de-sac*. These organisms have structural and physiological features rendering them unique tools for physiological and genetic analyses. The *Paramecium* cell is approximately 1000-fold larger than a typical animal cell, and by virtue of its large size, it requires more than a single diploid number of chromosomes to direct its activities. Ciliates have, therefore, evolved a novel structure, the *somatic* macronucleus, which is polyploid, containing many copies of the genome. It represents the soma—the DNA fraction that is actively transcribed to direct cellular activities. Each cell also contains two *germinal* micronuclei, which are diploid and function as the germ line. The micronuclei are transcriptionally inactive.

Cell division is accompanied by precise mitotic division of the micronuclei but amitotic division of the macronucleus (Figure 13.1). In the latter process, the numerous chromosomes within the macronucleus replicate, and then they separate into two roughly equal amounts in a random fashion as the macronucleus elongates, constricts, and separates into two nuclear regions. Over time during asexual growth, individual cells will segregate specific chromosomes and consequently possess less than the complete complement of macronuclear genetic material.

After conjugal union of two *Paramecium* cells of opposite mating type, the two micronuclei of each partner undergo two reductive meiotic divisions while the macronucleus is degraded. Seven of the eight haploid micronuclei then disintegrate, and the remaining nucleus undergoes a single mitotic division to give two identical copies of the haploid genome (Figure 13.2). One of these haploid nuclei, the male or sperm nucleus, migrates to the sexual partner while the other, the female or stationary nucleus, remains in the parental cell. Thus, each exconjugant receives a haploid nucleus from its mate so that the two daughter cells are genetically identical. After nuclear fusion, which reestablishes the diploid state, the nucleus undergoes two sequential mitotic divisions, and two of the daughter nuclei develop into macronuclei. The first cell division is accompanied by mitotic division of the two micronuclei, but the macronuclei do not divide. Instead, they migrate to opposite poles of the cell so that cellular constriction

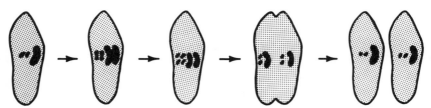

Figure 13.1. Binary fission (asexual reproduction) in *Paramecium*. The two micronuclei divide by mitosis while the macronucleus separates amitotically into two macronuclei after chromosomal replication.

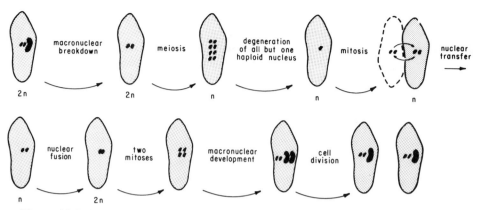

Figure 13.2. Nuclear events accompanying conjugation in *Paramecium*. Two cells of opposite mating type form a conjugation bridge, and the nuclear events depicted occur in each partner.

reestablishes the normal chromosomal complement of one macronucleus and two micronuclei per sister cell. A subclone containing cells in which the macronuclei descended from the same new macronucleus is referred to as a *caryonide*. Thus, there are two *sister caryonides* from each fertilized cell (Figure 13.2).

Paramecium is also capable of self-fertilization, a process called *autogamy*. Autogamy is very similar to the conjugal process depicted in Figure 13.2 except that a single cell is involved, so that nuclear transfer cannot occur. Instead, the two germinal haploid nuclei fuse, resulting in a diploid organism that is homozygous at all genetic loci.

Tetrahymena species are fundamentally similar to those of *Paramecium*, but they differ in a few respects. For example, they exhibit a distinct and characteristic morphology, are incapable of autogamy, and have only one micronucleus per cell. As in *Paramecium*, the macronucleus functions to direct cellular activities while the micronucleus functions exclusively to maintain the germ line.

Macronuclear differentiation in *Tetrahymena* has been studied in some detail. Following conjugal transfer and fusion of the germinal haploid nuclei, two of the four mitotically generated diploid nuclei are destined to become macronuclei and undergo repeated chromosomal replication so that within less than a generation the diploid (2n) number of chromosomes is increased to 64n. Then within the next 100 generations, this number gradually decreases to 45n. In *Paramecium*, the macronucleus has 800 times more DNA than its haploid set. Within the macronucleus, adenine bases become methylated, and there is evidence that methylation may be restricted to regions of the genome that are transcribed. Methylation of the germinal DNA does not occur. Of further interest is the fact that 10–20% of the sequences found in the micronucleus are apparently absent from the macronucleus. One DNA sequence that has been cloned from the micronucleus is clearly lacking from the macronucleus of the same cell. Other changes associated with macronuclear development involve tandem replication of the ribosomal RNA genes until the single copy in the original micronucleus gives rise to about 10,000 copies in the macronucleus, and translocation of a six base pair repeat (CCCCAA) occurs such that the flanking sequences change. While the mechanisms and developmental significance of these observations are not known, they clearly show that the macronucleus undergoes irreversible structural changes during development.

Control of Gene Expression in Ciliates

The study of ciliate genetics is complicated by the fact that numerous traits are not controlled in a Mendelian fashion. For example, in Table 13.1, four cellular traits are compared in two strains of *Paramecium tetraaurelia* before and after conjugation. It can be seen that although both exconjugants are genetically identical, as discussed above, all four traits listed including mating type, cell surface ciliary antigen, sensitivity to the toxin, paramecin, and the pattern of ciliary rows on the cell body are subject to cytoplasmic control. Distinct control mechanisms appear to operate for each of these traits. For example, paramecin sensitivity is controlled by particles in the cytoplasm of the cell, while the pattern of the ciliary rows is controlled by elements associated with the cell surface.

Control of ciliary antigen expression has been studied in some detail. More than 12 different ciliary antigens have been identified, and only one of these can be expressed on the cell surface at any one time. Expression is controlled genetically since different strains have different arrays of antigenic potential, but determination of the locus to be expressed is influenced by environmental factors such as temperature. Consider a conjugal mating between two strains of *Paramecium* of ciliary antigen genotypes *AAbb* and *a'a'B'B'*, respectively. In the first strain the *A* antigenic locus is expressed while the *b* locus is silent. In the second strain the *B'* locus is expressed while the *a'* is silent. If gene expression were under nuclear control, one would expect both exconjugants to express both *A* and *B'*, but not *b* or *a'*. The observation, however, is that the exconjugant

Table 13.1. Inheritance of Four Sets of Cellular Specificities in *P. tetraaurelia,* stock 51

Property	Strain 1	Strain 2	Exconjugant 1[a]	Exconjugant 2[a]
Mating type	VII	VIII	VII	VIII
Ciliary antigen expressed	A	B	A	B
Paramecin	resistant	sensitive	resistant	sensitive
Pattern of ciliary rows	normal	inverted	normal	inverted

[a]Strains 1 and 2 conjugate to give rise to exconjugants 1 and 2, respectively, which are of the same genotype. Because exconjugant 1 retains the properties of strain 1, and exconjugant 2 retains the properties of strain 2, it must be concluded that the traits listed are subject to cytoplasmic control.

derived from the first strain expresses both A and A' but not b or b', while the second exconjugant expresses B and B' but not a or a'. If cytoplasmic exchange occurs during conjugation, the two exconjugants assume the same serotype, which is environmentally determined. This phenomenon has been termed *interlocus repression* and was discussed in Chapter 11 as it applies to animal cells.

Another genetic phenomenon that has been well characterized in the ciliated protozoans is termed *allelic exclusion*. Following conjugation between two strains that are respectively of the AA and $A'A'$ genotypes at a particular locus, the two exconjugants initially express both the A and A' genes. After several generations, some of the progeny begin to show quantitative diminution in the expression of one of the alleles, relative to the other, and subsequently expression of that allele is extinguished altogether.

Allelic exclusion is stable, even if both alleles are turned off by intergenic repression, and may be due to loss of the genetic material coding for one allele or the other during amitotic nuclear division in the macronucleus. The phenomenon can be directional for specific alleles, but all alleles appear to assort with similar rates. Thus, phenotypic assortment may be possible at every locus, given sufficient time.

Mating Type Determination in Ciliates

Each pair of mating types can be thought of as a subspecies within a species complex. Within a species complex of either *Paramecium* or *Tetrahymena* there may exist numerous pairs of mating types that are fertile only with each other. For example, in *P. aurelia* there are 14 known subspecies designated *P. primaurelia, P. diaurelia, P. triaurelia, P. tetraaurelia, P. pentaaurelia*, etc., and consequently there are 28 known mating types. While in many ciliates, a cell of a given mating type can mate only with a single cell type of opposite mating type, in other species, conjugation with a multitude of mating types is possible. Presumably this difference results from the presence or absence of genetic exclusion factors that control mating among related subspecies.

The nature of mating type transmission (inheritance) varies widely among these organisms. Mating type can be genetically regulated and genetically stable (synclonal uniformity), or it can be genetically regulated, but epigenetically unstable (synclonal variation). Further, nuclear differentiation of mating type can be determined randomly, probably by a mechanism that is dependent on environmental conditions such as temperature (the epigenetic A pattern), or it can be regulated cytoplasmically, and thus determined by the cytoplasmic state of the parental cell. In this second case, the genotype merely determines an array of potentialities that are cytoplasmically controlled (epigenetic B pattern). Finally, in some ciliated protozoa, mating type can show diurnal periodicity. In this situation, mating type interconverts as a part of the cellular circadian rhythm. Consequently, the cell will be of one mating type during part of the day and the other mating type for the remainder of the day. In one *Paramecium* species, a single recessive allele abolishes the rhythm, and mating type determination assumes the epigenetic A pattern (random inheritance).

Mating types within each of the 14 subspecies of the *P. aurelia* species complex are designated either "O" or "E." All O cells within the species complex are homologous, as are all E cells. Two alleles at a single locus, the normal, dominant mt^+ allele and the mutant, recessive mt° allele, control the possibility for macronuclear determination. While a cell carrying an mt^+ allele can be determined for either the E or O mating type, mt°/mt° homozygotes can express only mating type O. Because attempts to produce mutants restricted to mating type E have failed, it is hypothesized that the O state represents a deficiency (inactivity) in a genetic regulatory component that is present (active) in the E state.

Mating type is determined randomly (epigenetic A pattern) during new zygotic macronuclear development following conjugation or autogamy and is almost always passed unaltered from mother to daughter cell during asexual reproduction. Because offspring of autogamy, which possess micronuclei that are homozygous at all loci, can be of either mating type, the genetic information for both mating types must be included within each haploid wild type genome. In some species of *Paramecium* in which two macronuclei develop within a single cytoplasm, each of the two developing macronuclei is determined independently. Therefore, macronuclear differentiation of mating type cannot be cytoplasmically controlled. However, the probability that the mating type will be E is dependent on the temperature during macronuclear development as well as on other environmental conditions. For example, the frequency of mating type E determination increases with increasing temperature. This observation leads to the possibility that a genetic regulatory constituent synthesized within the macronucleus and encoded or activated by the mt^+ gene determines mating type E. The absence of this substance gives rise to mating type O as occurs in mt° homozygotes. This substance must not be capable of free diffusion from one macronucleus to the other. Higher temperatures during macronuclear differentiation may promote either replication or expression of the mt^+ gene, or at higher temperatures the mt^+ gene may promote replication or expression of other genetic elements that are essential to development of mating type E. These observations lead to the suggestion that mating type E determination results from the development of structural characteristics of the macronucleus that are lacking in mating type O.

One species of *Tetrahymena* exhibits similar mating type determination behavior with one interesting complication. A gene, designated mt^A permits expression of mating type E and O while an alternative allele, mt^B, permits expression of two different mating types, E' and O'. E can mate with O but not O', while E' can mate only with O': In the mt^A/mt^B heterozygote, all four mating types can be expressed. This result establishes that a single gene can control fertility in ciliated protozoa, and that differences between two different subspecies within a species complex may have arisen from a single genetic alteration.

Many ciliated protozoa exhibit multiple mating types, some of which may be interfertile while others are not. In one species 7 interfertile mating types were found, while in another 48 mating types were reported. It is therefore clear that in ciliates a species designation becomes arbitrary and almost meaningless. Fertility barriers must arise and disappear by simple mutation with relatively high frequency.

Irreversible Mating Type Determination by DNA Excision

The above discussion leads to the conclusion that mating type determination in ciliated protozoa probably occurs by several distinct mechanisms. Only in a few cases have sufficiently detailed data been collected to allow formulation of a molecular mechanism. One such example is that of *Tetrahymena thermophila*. In this organism there are seven mating types, designated I–VII, each of which is capable of conjugation with cells of any of the other six mating types, but not with cells possessing the same mating type. Thus, mating in this organism is governed by a self-incompatibility system. Once established during macronuclear development, a mating type is passed on to all progeny during asexual reproduction. Changes in mating type during vegetative growth have never been reported. This obsevation leads to the possibility that in *T. thermophila,* mating type determination involves an irreversible, hereditary structural alteration of the macronuclear DNA. This alteration may occur at the mating type locus, *mt,* and involve irreversible excision of DNA sequences found in the germinal micronuclei.

Consistent with this suggestion, mating type determination in *T. thermophila* exhibits some characteristics typical of mutation. For example, the stability, clonal inheritability and apparently random selection of mating type owing to some type of chromosomal change are all characteristic of mutational events. On the other hand, the high frequency of interconversion, the fact that there are only seven alternative mating states that are available for selection and that all of these are fully functional when expressed argues against a typical mutational event. In these respects mating type determination more closely resembles an epigenetic, developmentally controlled event. These observations suggest that the requisite DNA structural alterations are catalyzed by a specific enzyme system acting on a specific region of the ciliate genome, the *mt* locus.

Based on the observations and reasoning cited above, a molecular model of mating type determination involving DNA splicing and deletion has been proposed. In this model it is assumed that the seven mating type genes are oriented in a tandem array at the *mt* locus (see Figure 13.3). This linear "cluster" of mating type genes is preceded by a common promoter (CP), which permits initiation of transcription. Each mating type gene is assumed to carry a proximal recognition sequence (*r*) which is the substrate for the mating type "excisase" enyzme, responsible for excision of DNA segments. The excisase is nonspecific with respect to mating type because it only recognizes the *r* sequence, common to all mating type genes. The excisase may bind to two adjacent r segments, delete the intervening DNA sequence, and then rejoin the cut ends, regenerating a new functional r sequence derived from two separate r sequences (Figure 13.3).

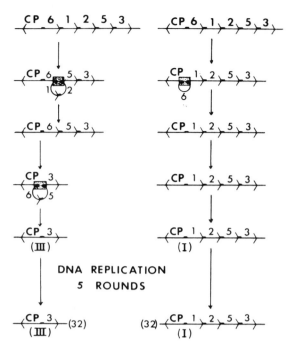

Figure 13.3. Depiction of a possible mechanism for mating type determination during macronuclear development in *Tetrahymena thermophila*. The *mt* region of the chromosome is depicted where CP is the common promoter, to the left, with the different mating type genes, to the right, designated by number. The thicker DNA segments are the *r* segments recognized and acted upon by the mating type excisase (see text). Dotted box: Mating type ligase, bound to two *r* segments. Left bracket: Initiation signal for transcription and/ or translation of the mating type polypeptide. Right brackets: Termination signals for transcription and/or translation of the mating type polypeptide. Only the DNA between the left bracket and the first right bracket can be expressed. From E. Orias, *Devel. Gen.,* 2:185–202 (1981), reprinted with permission.

The excisase is then released from the DNA, and the excised fragment is lost, either because it is degraded, or because it cannot be replicated. Implicit in this model is the assumption that the common promoter (CP) allows transcription of only that mating type gene which is adjacent to it. Possibly in the germ line, one mating type gene is already adjacent to the common promoter sequence. If so, it would be expected to be expressed in the absence of an excision event. Finally, the presence of multiple *r* sequences permits multiple excision events, but it is assumed that susceptibility to excision can only occur during a brief period of macronuclear development.

　　Because this model exhibits similarities to the postulated mechanism of immunoglobulin synthesis in higher animals, the latter process will be discussed in the next section. It is possible that DNA excision represents one of several strategies available to an evolving developmental system for effecting irreversible change. It must be noted, however, that such a mechanism can only be postulated when soma and germ line are maintained separately as is true of animals, plants and ciliated protozoa, and when the developmental process is completely irreversible.

DNA Excision in Immunoglobulin Synthesis

Immunoglobulins consist of heavy and light polypepide chains, each of which has a variable N-terminal region and a constant C-terminal region (Figure 13.4). There is a single class of heavy chains and two classes of light chains, termed λ and κ. Three gene clusters code for immunoglobulin chains: one which includes genes for heavy chain constant and variable regions, a second which consists of genes for λ light chain constant and variable regions, and a third comprised of genes for κ light chain constant and variable regions. Amino acid sequence analysis of all three classes of immunoglobulin chains (heavy, and both λ and κ light chains) reveals that a given variable region can coexist within a single polypeptide chain with different constant regions, and a particular constant region can likewise be associated with any one of several variable regions. It has been suggested that synthesis of an immunoglobulin polypeptide chain results from the fusion of the two genes coding for the selected variable and constant regions. Such a mechanism would be expected to generate a large diversity of antibodies.

Genetic and structural analyses of cloned mouse DNA have revealed that prior

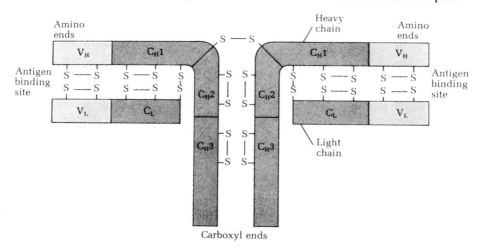

Figure 13.4. Structure of a typical immunoglobulin (IgG) molecule. The molecule consists of two light chains (L) (either κ or λ) and two heavy chains (H) joined by disulfide bonds. The constant (C) and variable (V) regions of the chains are indicated. Carbohydrate chains are attached to the heavy chains. (Amino termini and carboxy termini are indicated.) From A.L. Lehninger, *Principles of Biochemistry,* Worth Publishers, New York, 1982, figure 30-16, p. 927, reprinted with permission.

to immunogobulin differentiation, during early embryogenesis, the V and C genes are separate and nonadjacent, but that these genes are joined during embryological development. Figure 13.5 shows the possible arrangement of the immunoglobulin genes for one of the three classes of polypeptide chains (i.e., for the heavy or the λ or κ light chains) both prior to and subsequent to the excision event. The fusion of a variable region gene with a constant region gene to give a mature immunoglobulin gene results from removal of the DNA between these two genes, bringing them adjacent to each other. This results in the permanent loss from the cell of the structural genes for the variable and constant regions that lie between the variable and constant genes that were selected for expression. Kinetic hybridization studies have confirmed this suggestion and allowed tentative ordering of the genes coding for the constant regions of the heavy chains. The proposed gene order is: $(V_1, V_2, V_3 \ldots V_n)$-$\mu$-$\gamma_3$-$\gamma_1$-$\gamma_{2b}$-$\gamma_{2a}$-$\alpha$, where μ and α code for the constant regions of the heavy chains of IgM and IgA type immunoglobulins, and γ_1, γ_{2a}, γ_{2b}, and γ_3 represent genes coding for the constant regions of the different IgG immunoglobulin subclasses. Thus, the analyses carried out on immunoglobulin structural gene differentiation imply a programmed biosynthetic developmental mechanism involving DNA rearrangements concomitant with irreversible gene excision.

Aging and Programmed Death in Ciliates

Multicellular organisms exhibit mortal characteristics: They pass through developmental stages leading from immaturity to a sexually mature state followed by senescence and death. Somatic cells isolated from the body also exhibit the property of mortality, and they normally die after a number of cell divisions in tissue culture. It therefore has been suggested that mortality is a somatic cell property rather than an organismal or environmental characteristic. Mortal characteristics, resulting from inherent cellularly controlled genetic programs, are typical of ciliated protozoa. A simplified program of cellular differentiation is illustrated in Figure 13.6. Following nuclear reorganization, resulting from either

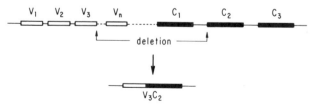

Figure 13.5. Proposed mechanism for the irreversible differentiation of immunoglobulin-synthesizing cells employing a mechanism that involves fusion of two nonadjacent but nearby genes. The abbreviations are as follows: V_1, V_2, V_3, $\ldots V_n$ = separate genes for distinct variable regions; C_1, C_2 $\ldots C_n$ = genes for distinct constant regions; V_3C_2 = a mature gene encoding an immunoglobulin polypeptide.

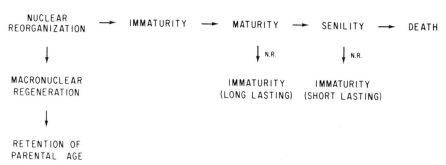

Figure 13.6. Schematic representation of the irreversible developmental program through which many ciliates pass. Nuclear reorganization by conjugation or autogamy is followed by a period of immaturity during which mating is prohibited. Acquisition of sexual maturity is a genetically determined and biochemically controlled event. Senility (decline) may result in a breakdown of genetic repression mechanisms and may be accompanied by old age selfing or sterility, depending on the species and strain. In the absence of nuclear reorganization (N.R.) death ensues, possibly as the terminal programmed event. Also as illustrated in the figure, the duration of the immaturity period may be influenced by the age of the conjugating parents.

conjugation or autogamy, ciliates pass through an immature period during which mating cannot occur. The duration of this period, preceding acquisition of sexual maturity, is genetically and physiologically determined. After a number of additional cell divisions, determined by the genetic composition of the cell, the mature cell passes through a period of decline or senility. If nuclear reorganization following conjugation or autogamy does not occur, the cell will eventually die.

Figure 13.6 also illustrates another interesting feature. The period of immaturity is influenced by the age of the parental cell undergoing nuclear reorganization. If a cell conjugates or undergoes autogamy shortly after becoming sexually competent, the period of immaturity of the exconjugant will be long lasting. On the other hand, if an older, even senile cell undergoes nuclear reorganization, the immaturity period is short lived. Thus, the exconjugant cell possesses a "memory" system for recalling the age of the parent.

The time to sexual maturity is a genetically adjustable interval in all well-characterized ciliated protozoa, and the temporal control settings can vary greatly. In some ciliates, the immaturity period is less than ten generations, and a few display no immaturity at all. In others it may last more than 200 generations. Yet the duration is normally fairly constant for any one inbred strain of ciliates. Variants of some ciliates with shorter immaturity periods (about 20 generations in *Tetrahymena pyriformis* compared with 50 for the normal strain) arise with fairly high frequency. These *early maturity* variants have been shown to breed true and can result from dominant genetic alterations at several distinct loci. While it is not known why the frequency of mutation to early maturity is high, nor what the genetic alteration entails, it seems reasonable that a relatively simple control mechanism involving one or a few central regulatory proteins might be involved.

Relevant to this suggestion, recent studies have led to the isolation of a protein that appears to control the duration of the immaturity period. This protein, termed *immaturin,* is a heat labile protein of a molecular weight near 10,000. It is found in the cytoplasm of immature *Paramecium* cells but not in that derived from sexually mature cells. When immaturin is injected into mature cells, mating activity is suppressed. Variations in the amounts of immaturin present in the cytoplasm during different stages of immaturity suggest that the protein controls the duration of the period by repressing the sexually active state.

Mating ability in ciliates has been shown to depend on the presence of complementary, sex-specific, cell-recognition molecules (mating substances) located on the ciliary membranes. The presence and specificities of the mating substances are controlled genetically. Since immature cells are incapable of sexual agglutination, immaturin possibly inhibits expression of mating substance activity. Although the mechanism of action of immaturin is as yet unknown, it may, for example, prove to be a transcriptional repressor of genes encoding mating-specific proteins, including the mating substances.

Macronuclear differentiation in ciliates consists in part of a program of allelic (phenotypic) assortment. Assortment of serotypes and enzymes occurs at different times during development, as measured by the number of cell fissions (Figure 13.7). In fact, it seems that most if not all genetic loci exhibit somatic assortment when heterozygotes are analyzed. The time of assortment is locus specific, but once assortment is initiated, the rate is similar for many different loci. This rate is independent of the allele being expressed, and occurs even if neither allele is being expressed, as a result of intergenic repression. Because assortment at any one locus is initiated at a specific time during differentiation, these events, taken collectively, comprise a "calendar" of the developmental program.

Events comprising the calendar could either occur independently of one another or be interdependent. In the former, but not the latter case, altering the timing of one event would be expected to change the time at which other events occur. As shown in Figure 13.7, early maturity mutants are altered only with respect to the acquisition of mating capacity. Thus, the temporal program does not involve a master calendar that controls all events in an obligatory sequence. The same conclusion was reached for *Drosophila* development as discussed in Chapter 1.

In many ciliated protozoa, senescence and death represent an apparently obligatory part (the terminal part) of the developmental program. For example, *Paramecium tetraaurelia* cells decline in vigor after about 150 generations. Subsequent progeny lose their capacity for photorecovery from ultraviolet irradiation damage, and show decreases in (1) the rates of DNA and RNA synthesis, (2) DNA template activity, (3) the amount of macronuclear DNA, (4) the number of food vacuoles, (5) nucleolar density and (6) number of polysomes. There are also decreases in the fission rate and the survival rate after autogamy. Senescence is accompanied by the breakdown of repressive mechanisms established during

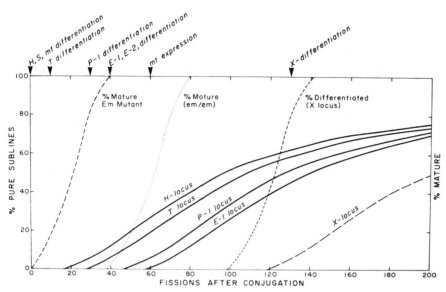

Figure 13.7. Semischematic depiction of the *Tetrahymena* "calendar." The maturation curve for the wild type (*em/em*) strain (dotted line) is shown shifted to an earlier time for the early maturity (*Em*) mutant (dashed line to the left). The distribution of other differentiative events (locus fixations) has not been examined carefully, but is assumed to be similar to that in maturation curves. The mean times of these differentiations are indicated on the temporal map across the top axis. The assortment of the mosaic macronuclei generated by these differentiations yields, after a lag, the phenotypic curves shown in solid lines. Most differentiations occur within the first 50 fissions post-conjugation, but assortments continue for long periods of time. From D.L. Nanney in *Mechanisms of Ageing and Development* 3:81–105 (1974), Elsevier Sequoia S.A., The Netherlands, reprinted with permission.

development. An example of this sort of behavior is provided by *Euplotes crassus,* which exhibits the phenomenon of old age selfing. During development, a single mating type is selected by expression of one of several alleles at the mating type locus. These alleles manifest "peck order" dominance relationships so that only one is selected for expression. In old age, however, heterozygotes become capable of selfing because of a breakdown in the dominance relationships so that mutually exclusive mating type alleles show irregular oscillatory expression. Thus, after a number of cell divisions, mating type will change so that cells of opposite mating type will be present in a clone that was originally of a single, self-sterile mating type. Since not all heterozygous clones exhibit old age selfing, it appears that this trait is determined by polymorphic genes.

In contrast to *P. aurelia* and *E. crassus, T. pyriformis* may not possess an obligatory genetic program for somatic senescence. Death is not necessarily a scheduled event in the life cycle. However, aged *Tetrahymena* may lose the capacity for nuclear rearrangement, and short-lived variants of *Tetrahymena* are known, which show an age-related onset of cell divisional disorganization that is followed by clonal death. Long-lived variants of *Paramecium* have also been reported. It appears that somatic senescence in ciliates is a matter of evolutionary choice. The organisms can, but need not be programmed for a finite life span. The variations in the lengths of the life cycle in different strains of *Tetrahymena* and *Paramecium* suggest that during recent evolutionary history a program including senescence and death may have been favored by certain ciliates that were subject to a certain set of environmental conditions, while indefinite growth may have better suited other ciliates that were exposed to different evolutionary pressures. It might be guessed that cellular mortality first evolved in this way as a matter of natural selection, and that only after the advent of multicellularity did somatic mortality become fully obligatory for species survival.

What evolutionary pressures gave rise to programmed death? Older organisms may be *genetically* dead in the sense that they are incapable of conjugation, recombination, and rapid evolutionary change. These cells nevertheless occupy the same ecological niche as younger, conjugation-competent members of the species. They therefore compete with them for nutrients, space, and favorable conditions. Competition limits the survival chances of the younger, more adaptable members of the species which, because of recombination, should be more capable of responding to environmental change by exhibiting beneficial genetic variation. In this regard, it is interesting to note that amicronucleate ciliates have been isolated and maintained by continuous asexual growth for several decades. The fact that these cells are rare in nature argues that a sexual mode of reproduction must be advantageous to species survival. In a diploid organism, sexuality allows the masking and maintenance of lethal recessive alleles within the population (short-term advantage) and provides for maximal genetic adaptability in response to environmental change, as a result of recombination of potentially advantageous alleles among outbred strains (long-term advantage). This argument therefore provides a rationale for the evolution of mortality as part of a linear sequence of cellular differentiation.

Selected References

Haga, N. and K. Hiwatashi. A protein called immaturin controlling sexual immaturity in *Paramecium, Nature 289:*177 (1981).

Hayflick, L. Current theories of biological aging, *Fed. Proc. 334:*9 (1975).

Lockshin, R.A. and J. Beaulato. Programmed cell death, *Life Sci. 15:*1549 (1974).

Nanney, D.L. Ciliate genetics: patterns and programs of gene action, *Ann. Rev. Gen.* 2:121 (1967).

Nanney, D.L. "Aging and Long-Term Temporal Regulation in Ciliated Protozoa. A Critical Review" in *Mechanism of Ageing and Development,* 3, Elsevier Sequoia S.A., Lausanne, 1974.

Orias, E. Probable somatic DNA rearrangements in mating type determination in *Tetrahymena thermophila:* a review and a model, *Devel. Gen.* 2:185 (1981).

Schneider, E.L., ed. *The Genetics of Aging*. Plenum Press, New York, 1978.

Saunders, J.W. Death in embryonic systems, *Science 154:*604 (1966).

Sonneborn, T.M. Genetics of cellular differentiation: stable nuclear differentiation in eucaryotic unicells, *Ann. Rev. Gen. 11:*349 (1977).

Todaro, G.J. and H. Green. Quantitative studies of the growth of mouse embryo cells in culture and their development into established lines, *J. Cell Biol. 17:*299 (1963).

Wright, B.E. and P.F. Davison. Guest editorial: mechanisms of development and aging, *Mech. Ageing Develop. 12:*213 (1980).

CHAPTER 14

Conclusions and Perspectives

To see a World in a grain of sand,
And Heaven in a wild flower,
Hold Infinity in the palm of your hand,
And Eternity in an hour.

Blake

In this volume we have attempted to integrate our knowledge of biological differentiation by using an evolutionary framework and examining related processes in organisms from divergent phylogenetic lines. Similar patterns of cyclic and linear differentiation were found to repeat themselves in organisms of surprisingly different cell structure and organismal complexity such as *Bacillus* and *Saccharomyces, Streptomyces* and *Neurospora, Myxococcus* and *Dictyostelium,* and *Anabaena* and *Hydra* (Chapters 4 and 5). This observation led to the suggestion that the fundamental molecular mechanisms controlling programs of differentiation may have existed prior to the divergence of eukaryotic microorganisms from prokaryotes. Alternatively, genetic exchange across phylogenetic lines may have occurred repeatedly during evolutionary history (Chapter 3). Moreover, we suggested that the advent of linear programs of differentiation, first described for blue-green bacteria and subsequently at a much greater degree of complexity in higher eukaryotes, may have necessitated the advent of programmed death (for elimination of genetically less competent individuals) and sexual conjugation (for the maintenance of species continuity) (Chapters 1, 3, and 13). Assuming this to be true, programs of differentiation necessarily coevolved with sex and programmed death, for these three phenomena represent essential components of a single developmental process. Similar fundamental mechanisms can therefore be postulated for the control of differentiation, sex, and programmed death.

Embryological development gives rise, from a single cell, to a multicellular organism consisting of two fundamental cell types: mortal somatic cells, and immortal germ cells (Chapter 3). While the former are believed to be programmed to die when their developmental programs run out, the latter cells retain their primordial capacity for eternal life because they have not become entrained into an irreversible, linear program of differentiation. An appreciation of this fundamental difference between somatic and germ cells resulted from detailed studies of cancerous transformation in mammals. While somatic cell transformation always depends on genetic alterations that partially destroy the program and release growth regulatory constraints, malignancy arising from a primordial germ cell depends on neither alteration because the germ cell was never committed to a program. Hence, stem cells within a teratocarcinoma retain totipotency for differentiation into all cell types of the body and possess the ability to participate in normal embryogenesis (Chapter 3).

Examination of regulatory mechanisms controlling cell physiology, sex, and differentiation have led to an appreciation of three fundamentally different molecular mechanisms. These mechanistic types can be classified as chemical, electrical, and macromolecular. Small chemical morphogens probably control morphogenesis in such divergent organisms as *Anabaena, Hydra,* and *Dictyostelium* (Chapter 5) while pheromones clearly regulate sexual activities in simple bacteria, yeast, and insects, to cite a few examples (Chapter 9). On the other hand, bioelectric activities mediate germination of fertilized eggs in brown algae, and possibly of germ tube formation in fungi, and of "shmoo" formation in yeast as well (Chapters 2 and 9). One function of transcellular electric fields and currents may be to allow the polarized deposition of cell surface biosynthetic precursors (Chapter 2). Electrical activities also play important roles in öocyte nutrition (*Hyalophora,* Chapter 2) and in initiation of embryogenesis following sperm–egg fusion (sea urchins, Chapter 10). Finally, macromolecules, usually proteins, function to relay information from one cell type to another as well as to transmit signals from one intracellular site to another (Chapter 7). While it is useful to recognize the three types of regulatory processes that influence differentiation, it should also be noted that they may be interrelated and mutually dependent. A particular reception-signal transmission-response relay system may depend upon elements of all three mechanisms.

Bioelectric activities are regulated at the level of transmembrane transport. Several distinct mechanisms of transmembrane transport have been elucidated in recent years, and the transport systems that catalyze vectorial processes can be regulated by small chemicals, by ion gradients and electric potentials, and also by macromolecules (Chapters 6 and 7). Moreover, transport proteins may exhibit multiple functions, acting catalytically to modify their substrates chemically, serving as chemoreceptors, and regulating other cellular processes. The identification of multifunctional transport proteins with as many as four distinct catalytic activities has led to the likelihood that enzymes, permeases, receptors, and regulatory proteins have a common evolutionary origin. Thus, mechanisms of chemical, electrical, and macromolecular information transfer may be related not only in terms of overlapping functions, but also in terms of structure and evolutionary origin.

Just as a diversity of transport and regulatory mechanisms is utilized by living cells, numerous reception mechanisms are available. Virtually all cells recognize and respond to small molecules in their environment, and their responses must have evolved to enhance species survival. Chemoreception regulates cell movements, morphogenesis, and growth during embryogenesis in multicellular organisms and serves analogous functions in appropriate microorganisms (Chapter 8).

Cell–cell recognition is perhaps the most complex form of biological chemoreception yet studied. Living cells can distinguish self from nonself by homotypic and heterotypic adhesive processes, and they exhibit numerous highly specific responses to cell–cell contact. These responses include transcriptional regulation, control of cell motility, and alterations in the growth rate. While all three of these responses have been well documented, little is yet known about the molecular mechanisms by which they are elicited (Chapter 10).

Just as we are rapidly learning about the biochemistry of differentiation, so are we gaining insight into developmental genetic regulatory mechanisms. In Chapter 7 the regulation of gene expression by intracellular and extracellular chemicals, mediated by ligand-binding proteins, was discussed. In Chapters 4, 11, 12, and 13, five distinct "switch" mechanisms were considered for controlling gene expression. These mechanisms include (1) sequential sigma factor synthesis and a consequent temporal change in the specificity of RNA polymerase; (2) DNA methylation at regulatory sites that control the expression of adjacent structural genes; and (3) DNA rearrangements. DNA rearrangements affecting gene expression may occur by inversion of a segment of the nucleic acid (Chapter 11), by transposition of a DNA fragment from one part of the genome to another (Chapter 12), or by deletion or insertion of genetic material (Chapter 13). Those genetic rearrangement processes that are of developmental significance are probably catalyzed by specific enzymes. Thus, gene inversion would be catalzyed by a *flipase*, transposition by a *transposase* and excision by an *excisase*.

Theoretical considerations lead to the conclusion that any one of the five recognized mechanisms for switching genes on or off can occur in either a reversible or an irreversible fashion. We shall consider these possibilities one at a time.

In *Bacillus*, there is evidence for several different sigma factors that interact with RNA polymerase and regulate its operon specificity (Chapter 4). Let us assume that one sigma factor (Σ_1 in Figure 14.1) allows transcription of vegetative genes encoding proteins P_1, P_2, P_3. . . . These proteins are synthesized as long as a sufficient nutrient supply is available. Starvation conditions, however, elicit a signal that promotes synthesis of a new sigma factor (Σ_2). If this change in transcriptional expression is to occur reversibly, Σ_1 would either remain active or it would be reversibly repressed by the starvation signal. On the other hand, if activation of Σ_2 is to occur irreversibly, Σ_2 might activate a gene that encodes an inhibitor of Σ_1 (I_1), and I_1 would inactivate Σ_1. In this way no vegetative regulons would continue to be expressed. If it is further assumed that Σ_2 activates a gene that codes for a new sigma factor (Σ_3) and that Σ_3 activates transcription of an inhibitor of Σ_2 (I_2) as well as a new sigma factor (Σ_4), an irreversible program of gene expression would have been effected.

In Chapter 11, the recent evidence suggesting that DNA methylation can regulate gene expression was discussed. Methylation of cytosine residues at

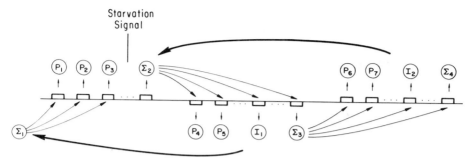

Figure 14.1. Proposed sequence of sigma factor-controlled transcription–translation events leading to an irreversible program of gene expression. The designations are as follows: Σ_1, Σ_2, Σ_3 . . . Sigma factors determining the operon specificities of RNA polymerase. P_1, P_2, P_3 . . . , proteins encoded by structural genes that are activated by the sigma factors. I_1, I_2 . . . , inhibitors or inactivators of the respective sigma factors. $\rightarrow$ activation; $\rightarrow$ inhibition.

specific GC regulatory sites, adjacent to certain genes encoding differentiation-specific proteins, decreases or prevents expression of that gene. The methylase apparently recognizes a methylated cytosine in the old complementary strand and catalyzes methyl transfer to the adjacent cytosine in the newly replicated DNA strand. Consequently, gene silencing by methylation can be inherited from generation to generation. Methylation of other regulatory sequences may activate adjacent genes as might be suggested for the ciliates (Chapter 13).

In Chapter 11, two gene inversion events were considered, flagellar phase variation in *Salmonella,* and G-loop inversion in the bacteriophage, Mu. In the former case, the invertible 890 base pair segment encodes the *flipase* that catalyzes inversion as well as the *H2* operon promoter. Because the flipase is synthesized regardless of the orientation of the invertible DNA segment, inversion is fully reversible. Similarly, G-loop inversion in phage Mu occurs reversibly. In this case, the structural genes for the tail fiber proteins are present largely within the 3000 base pair invertible segment, and the promoter as well as the flipase structural gene is outside this region. Again, because the flipase is always synthesized, regardless of the orientation of the G-segment, inversion occurs reversibly. If, however, the flipase gene (*gin*) were expressed only when the invertible DNA segment was in one of the two possible orientations, an irreversible inversion would occur. This is illustrated in Figure 14.2. In Figure 14.2A the flipase gene is part of an operon that is outside the invertible segment but under the control of a promoter present within the invertible segment. In this configuration, operon expression can switch from "on" to "off," but not back, if the flipase protein is lost as a result of degradation or dilution during growth. In Figure 14.2B, the opposite orientation is observed: The flipase gene is on the side of the invertible segment opposite to that of the operon. Consequently, flipase is synthesized only when operon expression is turned "off". This configuration allows irreversible switching from "off" to "on."

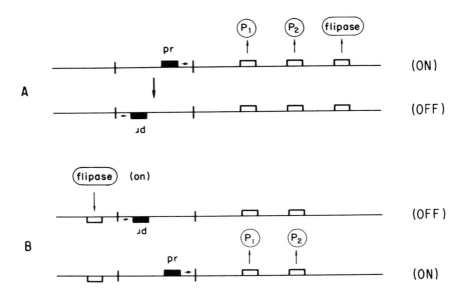

Figure 14.2. Two possible gene arrangements that allow genetic inversion to occur in an irreversible fashion. (A) The operon encodes proteins P_1 and P_2 as well as the flipase, which catalyzes the inversion event, and all three genes are transcribed when the promoter (*Pr*), within the invertible repeat, is adjacent to the first structural gene of the operon. In this configuration the inversion may occur irreversibly from the "on" to the "off" position. (B) The flipase structural gene is on the side of the invertible DNA repeat opposite to that of the operon. When the operon is silent the flipase is synthesized and vice versa. In this configuration, irreversible inversion from off to on can occur.

In Chapter 12, a transposition of genetic material from one of the *HM* loci to the *MAT* locus on chromosome III of *Saccharomyces cerevisiae* was shown to give rise to mating type interconversion. The enzyme catalyzing transposition, the transposase, is probably encoded by a gene on chromosome IV (the *HO* gene). Regulatory genes on the transposed DNA are responsible for activating and silencing dozens of sex-specific genes scattered throughout the yeast genome. If one of these regulatory genes were to control expression of the *HO* gene, such that this gene was silenced, or if the *HO* gene product were neutralized, an irreversible switch would be effected.

Finally, an excision event, as discussed in Chapter 13, would always be expected to be irreversible unless the excised genetic material were stored within the cell. Precise reinsertion could reverse the alteration in gene expression. It might also be noted that *insertion* of genetic material, the opposite of excision, or tandem duplications could also give rise to switch mechanisms that qualitatively or quantitatively influence either the expression or the potential expression of developmentally regulated genetic material.

While the molecular basis of developmental genetic control mechanisms is at present well established, both the diversity of mechanisms and the fine details of the processes need to be ascertained. It can be anticipated that research in the near future will provide clues to as yet unimagined mechanisms, as well as clarifying the molecular details of the concepts formulated within the recent past.

Nature hath no goal though she hath law.

Donne

The ancient covenant is in pieces;
man knows at last he is alone in the
universe's unfeeling immensity, out
of which he emerged only by
chance.
His destiny is nowhere spelled out,
nor is his duty. The kingdom above
or the darkness below. It is for
him to choose.

Jacques Monod

Index